RAIN OF IRON AND ICE

RAIN OF IRON AND ICE

THE VERY REAL THREAT OF COMET AND ASTEROID BOMBARDMENT

JOHN S. LEWIS

Helix Books

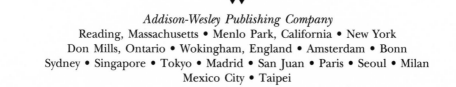

Addison-Wesley Publishing Company
Reading, Massachusetts • Menlo Park, California • New York
Don Mills, Ontario • Wokingham, England • Amsterdam • Bonn
Sydney • Singapore • Tokyo • Madrid • San Juan • Paris • Seoul • Milan
Mexico City • Taipei

Library of Congress Cataloging-in-Publication Data
Lewis, John S.
    Rain of iron and ice : the very real threat of comet and asteroid
bombardment / John S. Lewis.
      p. cm.
Includes bibliographical references and index.
ISBN 0-201-48950-3
1. Comets—Environmental aspects. 2. Asteroids—Environmental
aspects. 3. Impact. I. Title.
QB721.L42 1996
551.3'97—dc20   95–32915
                                                            CIP

Jacket design by Lynne Reed
Text design by Janis Owens
Set in 10.5-point Baskerville by Jackson Typesetting

1 2 3 4 5 6 7 8 9 10-MA-9998979695
First printing, December 1995

To my wife,

Ruth Margaret Adams Lewis

and

to my best friend

Peg,

who,

fortunately,

are one and the same person.

*Deux ou trois pages auraient suffi pour la Vérité;
les passions firent des Livres.*

ACADÉMIE ROYALE DES SCIENCES,
Paris. Histoire (1710)

TRANSLATION:

Two or three pages suffice to convey the truth;
it is passions that make books.

CONTENTS

ACKNOWLEDGMENTS

I am grateful for Melinda Hutson's helpful comments on the manuscript of this book, and especially to Dr. Guy Consolmagno for his careful review and detailed historical comments.

I am pleased to acknowledge the assistance of the following for giving permission for reprinting excerpts of other publications: *The New York Times,* for twenty brief quotations on pages 105, 126–130, 162, 166, and 172, the *Journal of Paleontology,* for quotations on pages 100 and 106, the *Astronomical Journal,* for the quotation on page 150, the *Journal of Geophysical Research,* for the quotations on pages 91 and 116, Dr. Brian Marsden, for the quotation on page 146, Ginn and Company for the quotation on page 138, and *Science* for the quotation on page 109.

The photograph of the Gibeon iron meteorite in the insert is courtesy of the Smithsonian Museum. The photograph of the meteor shower is courtesy of Dennis Milon. The two spacecraft images are courtesy of NASA. The image of the nucleus of Halley's Comet is courtesy of ESA. The Spacewatch asteroid discovery picture is courtesy of Tom Gehrels. The image of Phobos is courtesy of NASA.

RAIN OF IRON AND ICE

INTRODUCTION

It was a warm, clear afternoon in the capital. The bustle of metropolitan commerce and tourism filled the streets. Small sailing vessels dotted the sheltered waters within sight of the government buildings, riding on a soft southerly breeze. The Sun sparkled on the gentle swells and wakes, lending a luminous glow to the poppies and tulips nodding in the parks along the water's edge. All was in order. But suddenly the sky brightened as if with a second, more brilliant Sun. A second set of shadows appeared; at first long and faint, they shortened and sharpened rapidly. A strange hissing, humming sound seemed to come from everywhere at once. Thousands craned their necks and looked upward, searching the sky for the new Sun. Above them a tremendous white fireball blossomed, like the unfolding of a vast paper flower, but now blindingly bright. For several seconds the fierce fireball dominated the sky, shaming the Sun. The sky burned white-hot, then slowly faded through yellow and orange to a glowering copper-red. The awful hissing ceased. The onlookers, blinded by the flash, burned by its searing heat, covered their eyes and cringed in terror. Occupants of offices and apartments rushed to their windows, searching the sky for the source of the brilliant flare that had lit their rooms. A great blanket of turbulent, coppery cloud filled half the sky overhead. For a dozen heartbeats the city was awestruck, numbed and silent. Then, without warning, a tremendous blast smote the city, knocking pedestrians to the ground. Shuttered doors and windows blew out; fences, walls, and roofs groaned and cracked. A shock wave raced across the city and its waterways, knocking sailboats flat in the water. A hot, sulfurous wind like an open door into hell, the breath of a cosmic ironmaker's furnace, pressed downward from the sky, filled with the endless reverberation of invisible landslides. Then the hot breath slowed and paused; the normal breeze resumed with renewed vigor, and cool air blew across the city from the south. The sky overhead now faded to dark gray, then to a portentous black. A turbulent black cloud like a rumpled sheet seemed to descend from heaven. Fine black dust began to fall, slowly, gently, suspended and swirled by the breeze. For an hour or more the black dust fell, until, dissipated and dispersed by the breeze, the cloud faded from view.

Many thought it was the end of the world. . . .

Reconstruction of events in Constantinople, A.D. 472

Earth, like its sister planets, experiences an erratic but unceasing rain of comets and asteroids which has profoundly affected its geological

1

and biological history and which exposes its inhabitants to continuous risk of disaster. From time to time, enormous explosions affect the entire surface of Earth, excavating huge craters and making the biosphere briefly hostile to life. Much more frequently, small impactors have a devastating local effect similar to that of a multimegaton nuclear airburst.

Awareness of this danger dates back to the earliest human records. Mythology fairly drips with frightening allusions to cometary omens, celestial serpents, fiery dragons, mass disasters inflicted from above, entire cities destroyed by "lightning bolts," and the fall of alien materials from the skies. But myths, however often they may be based on actual events, have been filtered through countless generations of retelling, editing, and reinterpretation. Millennia after all eyewitnesses to these events have died, taking with them almost all of the circumstantial evidence of names, dates, places, and observed phenomena, the stories survive largely as cautionary tales, transmitted for their moral value and mythic grandeur. Their worth as firsthand accounts has been all but destroyed—but the fear, the horror, the raw emotional impact, have remained intact. The original eyewitnesses did not understand the phenomena they saw, and those who passed the stories down through the centuries neither understood nor observed any such events themselves. It is hard to doubt that the ancients witnessed countless events that are relevant to our story, but it is difficult to reconstruct these events from their lingering echoes in myth.

Although a general human awareness of cosmic bombardment can be traced back to ancient times, our ability to understand and predict these devastating events is of recent vintage. Fortuitously, a number of exciting but seemingly unrelated twentieth-century discoveries in many different fields of science have converged into a single vast drama; a kind of all-enveloping detective story in which we all are players. This story tramples traditional disciplinary boundaries and exposes time-honored philosophical principles to direct experimental tests. Our understanding of astronomy, geology, and biology is illuminated by this new insight: we see Earth's surface not in quasi-mystical terms as a uniquely sheltered refuge for life, but as a part of the fabric of the solar system, subject like other bodies to rare, cataclysmic change. The origin and fate of species are, we find, linked to events unknown to Darwin. The dominance of the human race, of mammals, of land life, the very existence of life on this planet may be consequences of our bombardment history.

It was not easy for the experts of earlier centuries to accept that Earth is constantly showered by rocky debris from asteroids and comets. Reports of meteorite falls by peasants were treated with ultimate disdain by "authorities" who derived their knowledge of nature not from observation and experiment, but from reading the dogma of Aristotle, Galen, Ptolemy, and the sages of the Church. It was only in the philosophical and political wake of the American and French Revolutions that such reports were sought out and analyzed by the real authorities: scientists who knew that their knowledge and understanding were imperfect, and who believed that both could be improved, perhaps even perfected, by diligent recourse to observation and critical analysis. The stories of several meteorite falls before 1800 are discussed in chapter 1 as illustrations of the great cultural change from the scholastic world of medieval Europe to the empirical world of the Age of Reason and of the changing of the guard at the Citadel of Truth and Traditional Wisdom.

If it was difficult to accept that kilogram-sized pieces of rock and iron could fall from the sky, it was vastly harder to contemplate mountain-sized impactors colliding with Earth at speeds of tens of kilometers per second. The first craters studied on Earth, which belonged to active volcanoes, called attention to themselves by means of spectacular eruptions. When other craters, showing different properties, were discovered, it was tempting to attribute them to the same familiar volcanic processes. Indeed, large craters with absolutely no sign of volcanic influence were disingenuously named "cryptovolcanic," meaning that all evidence of volcanic activity was *hidden*. It was only through extensive fieldwork and the recovery of numerous unambiguous meteorite fragments from Meteor Crater in Arizona, as recounted in chapter 2, that the door was opened to recognizing impact cratering as an important terrestrial process.

Early in the twentieth century, while American geologists were coming to grips with impact cratering on Earth, Russian scientists were puzzling over the astonishing event of June 30, 1908. Workers on the Trans-Siberian Railroad witnessed a brilliant daytime fireball burning across the sky to the north. The fireball passed over the northern horizon and had nearly faded from view when the sky was lit by a tremendous flash. A column of smoke and fire shot up above the horizon with astonishing violence. Even at great distances from the blast, many unusual effects were reported. The shock wave of a huge explosion traveled around the world, and dust from the blast lit the skies of western Europe for days afterward. But Russia was then in such a state of political chaos that no expedition could be mustered to search for the impact crater.

In 1927, when the political situation in Siberia had sufficiently settled down, an expedition was sent in to study the expected impact crater in the Tunguska taiga. But, as recounted in chapter 3, no crater was found. Instead, some two thousand square kilometers of boreal forest were devastated by fire and blast effects, attesting to an unprecedentedly powerful aerial explosion. What could possibly have caused such a violent blast? Why was no crater formed? How often do events of this sort, which leave no enduring geological record, happen on Earth? Many answers to these questions were proposed in the following years, but true understanding had to await the arrival of the atomic bomb.

From the time of the first nuclear explosion in New Mexico in 1945, and emphasized by the cruel fate of Hiroshima and Nagasaki, it was obvious that radiant heating by an extremely energetic fireball must have been the culprit at Tunguska. Experiments with atomic and hydrogen bombs in Nevada and Kazakhstan, described in chapter 4, led to a growing understanding of the physical and chemical effects of large explosions. Leaving aside those effects caused solely by nuclear reactions, very large nonnuclear explosions could be expected to produce powerful blast waves, sear the ground near them, and, if the explosion was on or near the surface, excavate a crater. The hot explosion fireball is very buoyant and rises rapidly, generating wind speeds of up to about a third of the speed of sound and forming the familiar mushroom cloud of rising, cooling gases and debris. Nitrogen in the air, heated to temperatures of many thousands of degrees in the fireball, partly burns to make nitrogen oxides, noxious gases that lend a red-brown color to the cooling fireball. Dust is raised by near-surface bursts and lifted to high altitudes by the mushroom cloud. Even modest-sized nuclear explosions can have effects detectable over intercontinental distances.

But full acceptance of the importance of large impacts on Earth had to await the exploration of the solar system by spacecraft. In the 1960s, beginning with the first unmanned missions to the moon, it was found that craters of all sizes were widely but unevenly spread over the entire lunar surface. The volcanic interpretation of lunar craters, popular among geologists early in the twentieth century, faded under the barrage of data from planetary scientists. Chapter 5 relates the discovery of heavy cratering on Mars and Mercury, and the application of these planetary data to interpretation of craters found on Earth by geologists. Application of space techniques to the study of Earth permitted the discovery and identification of many more terrestrial impact craters, and dating of materials returned from the Moon by both manned and unmanned missions permitted us to estimate the rate of cratering on both the Moon and Earth. By 1970 it was proven beyond doubt that impacts of comets and asteroids on Earth were of great importance over the span of geological time. Further, it became natural and timely to search for bodies in threatening orbits.

The new perspective provided by space exploration gave both scientists and the public a global view of Earth and an appreciation of its environment in the solar system. The recognition of both asteroids with orbits that crossed that of Earth and bodies that actually skimmed through Earth's atmosphere and escaped brought home the intimate association between cosmic and terrestrial events. Chapter 6 recounts the systematic search for Earth-crossing bodies, culminating in the application of modern sensor and computer technology to this problem in the Spacewatch program. The story of the discoveries by these search programs leads naturally into an extrapolation into the near future: How can we best search for threatening objects, and how many can we expect to find?

Another discovery from the early space age was the mechanism responsible for raising intense planet-wide dust storms on Mars. Computer modeling of the effects of dust on the Martian climate showed that large dust burdens in the atmosphere reflected much of the incident sunlight back into space, causing global cooling of the surface. In chapter 7 we see how atmospheric scientists tied together the story of dust storms on Mars and the effects of nuclear weapons to predict that a nuclear war would lift enough dust and soot into Earth's atmosphere to trigger a "nuclear winter." During the several months to one year that the dust remained in the atmosphere, temperatures in continental interiors would plummet, causing global crop failures and mass starvation.

Paleontologists studying the pattern of appearance and disappearance of species in the fossil record have long been aware of abrupt, devastating global extinction events occurring at the ends of geological ages. The greatest of these is at the end of the Permian era, and the second greatest marked the end of the Cretaceous era. In 1981 a team of physicists and geologists headed by Luis Alvarez of Berkeley discovered that a thin, global sediment layer that separates the end of the Cretaceous era (the last period of the age of dinosaurs) from the beginning of the Tertiary era (the start of the age of mammals) contained the unmistakable signature of an asteroid or comet impact. A number of metals, such as iridium, that are very rare in Earth's crust but common in meteorites, were found to be dramatically enriched in that layer. Further, as we discuss in chapter 8, the layer, which is dominantly composed of a very fine-grained clay, was found to contain tiny particles of minerals that had experienced extremely high shock pressures. The layer also contains a large amount of soot, and in some locations a generous admixture of tiny glassy beads called microtektites. The layer is at least a millimeter thick over the entire planet, enough to ensure an intense global nuclear winter, but is considerably thicker at some locations in the Americas. In Haiti it is found in association with, and painted on top of, a rubble layer tens of meters thick. Recently the "smoking gun" has been found: a huge 65-million-year-old crater over two hundred kilometers in diameter, buried under more recent sediments on the north shore of Mexico's Yucatán Peninsula. The end of the Cretaceous era is one of the most interesting and important markers in the geological history of Earth: it is the time of a devastating biological extinction event, in which the pattern of life on the continents and in the oceans changed dramatically in a geological instant.

It is reasonable to ask whether Earth and its twin, Venus, have experienced similar geological histories, but the dense, permanent cloud layer of Venus long frustrated man's efforts to study its surface. The *Magellan* radar-mapping mission was designed to penetrate the dense cloud layer and return detailed radar images of the surface geology. One of the most exciting results of this study of "Earth's twin gone wrong" has been the discovery that, unlike the Moon, Mercury, and even Mars, the number of craters on Venus does not increase rapidly toward smaller sizes: Not only are craters smaller than a few kilometers in diameter extremely rare, but there is a clear tendency for large craters to form pairs and clusters. Chapter 9 relates and interprets the discoveries of the *Magellan* mission regarding the final disintegration of comets and asteroids during high-speed entry into planetary atmospheres. The dense atmosphere of Venus serves as a cushion to not only protect the surface from kilometer-sized impactors, but also decelerate and capture gases released by comets and asteroids as they fragment and burn up. Atmospheric breakup, although less extreme on Earth than on Venus, is still an important factor here, giving us an insight into the nature of Tunguska-type impactors.

In the fall of 1992 a group of radar astronomers were examining Mercury at a time when the relative positions and orientations of Earth and Mercury provided an unusually clear view of the north polar region. Much to the astonishment of the observers, the radar image clearly showed a small, circular, highly reflective region on the floor of a large crater adjacent to the pole. The radar signature of this bright reflective material can be produced by only one known material: ordinary ice. Later observations of the other pole showed another small, bright cap there. Thus Mercury, scorched by the Sun at all other latitudes, almost miraculously has been able to produce and maintain two polar ice caps. Theoretical studies described in chapter 10 suggest that impacts of water-rich asteroids are the most likely source of ices on Mercury, and underline the importance of impacts as a source of atmospheric gases on Venus—and on Earth.

In the spring of 1993 came the astonishing news of the discovery of a peculiar comet-like body called Shoemaker-Levy 9. Photographs showed it to consist of twenty or more radiant pieces in a straight line, embedded inside an extensive bright cloud. A search for the distinctive spectral lines emitted by gases in the supposed comet turned up negative: the diffuse cloud is dust, not gas. As observations of this body accumulated, it was soon determined that Shoemaker-Levy 9 was actually in an extremely elongated elliptical orbit around Jupiter. This was quickly followed by the discovery that its orbit had taken it very close to Jupiter in July 1992, so close that the gravitational (tidal) attraction of Jupiter at the surface of Shoemaker-Levy 9 must have been significantly larger than its own surface gravity. In effect, it was pulled apart by Jupiter's powerful gravity. Even more astonishing, it was found that in its present orbit the entire "string of pearls" would impact on Jupiter in July 1994, causing a spectacular series of gigantic explosions extending over several days. Chapter 11 connects the most recent observations of Shoemaker-Levy 9 to other astronomical lore on the breakup of comets and asteroids, satellites of asteroids, crater clusters on Venus and Earth, and crater chains on Jupiter's giant icy satellites Ganymede and Callisto.

Given the evident importance of impacts on the atmospheric evolution of Venus, and the inevitability of oceanic impacts on Earth, it has become ever more urgent to understand the effects of large impacts on fluid (oceanic and atmospheric) targets. Applying current impact theory to Mars gives a startling result: impacts are capable of removing atmosphere from Mars and ejecting it into space at a speed greater than the planet's escape velocity. Impacts erode the

. . . .

atmosphere irreversibly. Over geological time it is likely that impacts have reduced the mass of the Martian atmosphere by a factor of one hundred. Impacts on Earth differ from those on Mars in that terrestrial impacts have a high likelihood of encountering an ocean. A violent explosion in the ocean will inevitably raise powerful waves, or tsunamis, that can run thousands of kilometers across deep ocean basins with trough-to-crest heights of a few meters and crest-to-crest lengths (wavelengths) of several kilometers. Chapter 12 describes how tsunami waves narrow as they encounter the shallows above continental margins, climbing to heights of hundreds of meters or even kilometers above sea level. Moderate-sized impacts, even in remote oceanic areas with low population densities, may have devastating and lethal effects half a world away.

The effects of comet and asteroid impacts are potentially damaging to life in general, and to human civilization in particular. In the long term, impacts with devastating global effects are very rare, happening in intervals of 10 million years or so. On the scale of a human lifetime, global catastrophes are improbable: the probability of such an event (an impact of 100 million megatons or more) happening during any given generation is less than 0.00001. But *locally* devastating impacts are vastly more common. A mere one hundred megatons can cause severe local destruction, and one-hundred-megaton explosions occur at an average rate of one per millennium. In the twentieth century alone, tens of megatons of explosive power have been liberated in the atmosphere by cosmic impacts. About three-fourths of these impacts struck the oceans and polar regions. The Tunguska event alone was about fifteen megatons, and several other multimegaton events probably occurred in remote, unobserved areas. What are the real hazards from these smaller, much more numerous impactors? Chapter 13 collects eyewitness reports on damage, injuries, and deaths from meteorites and impactor explosions over the past few centuries and discusses the reliability and completeness of the information at our disposal. It also summarizes reports of mysterious aerial explosions over metropolitan areas drawn from newspapers and magazines.

Pulling together all we presently know about the compositions, strengths, orbits, and numbers of comets and asteroids in near-Earth space, we can develop a computer simulation of bombardment activity for a typical century. By running this model for many centuries, we develop a sense not only of the magnitude of the threat to life and property, but also of the extreme randomness and variability of the natural bombardment process. Chapter 14 presents the results of ten different one-hundred-year runs, based on the most recent information. The effects on urban areas of blast waves, fire storms, and tsunamis are shown to be the most serious short-term ramifications of cosmic bombardment.

But the same rapidly developing technology that gives us the ability to identify, quantify, and assess the threat from comet and asteroid impacts also gives us some ability to counter these threats. Postimpact cleanup is ludicrously ineffective, but it may come as a surprise to some that there are technologically feasible and economically affordable forms of "collision avoidance" or "bullet deflection" that can protect us against most of the serious threats. The advantages and disadvantages, costs and benefits, and strengths and limitations of several proposed remediation schemes are discussed in chapter 15. It is clear that the threat can be reduced severalfold, but, in the final analysis, no defensive technology we can reasonably foresee and afford can guarantee residents of the universe a life of perfect safety. Interestingly, the greatest reduction of the impact hazard follows not from the negative impetus to avoid impacts, but from a positive motivation: the same space technologies developed to warn against impact threats can also be used to discover economically attractive bodies, both comets and asteroids, for use in support of human activities both in space and on Earth. Rather than being an endless drain on our pocketbooks, these same objects might prove to be sources of vast (and calculable) mineral and energy wealth. (This exciting story will be told in my next book, *Age of Iron and Ice: Civilization in Space.*)

1

STONES THAT FALL FROM THE SKY

By wil des dritten Friderich	In the reign of Frederick III
Geboren herr von Oestereich	Born Lord of Austria
Begt har in diss sin eigen landt	There fell in this his own land
Der stein des hie ligt an der wandt.	The stone that lies here against the wall.
Als man zalt vierzehnhuendert Jar,	As one counts 1400 years
Uff sant Florentzen tag ist war	On St. Florentius' day it was
Nuentzig und zwei umb mittentag.	In ninety and two around midday.
Beschach ein grusam donnerschlag	There was a great thunderclap
Drij zentner schwer fiel diser stein	Three hundredweight fell this stone
Hie in dem feld vor Ensisheim.	Here in the fields near Ensisheim.

SEBASTIAN BRANDT
Eyewitness account of the fall of the Ensisheim meteorite, November 1492

Long before modern times, long before there was such a thing as a university or a scientist, long before the invention of alphabets and the rise of literacy, our remote ancestors observed brilliant fireballs in the sky, heard powerful sonic booms, and saw occasional meteorites fall to the ground. Burdened with fewer institutions, there were fewer experts to challenge the evidence of the eyes and ears, fewer institutional prerogatives to maintain, less incentive to fit facts to theory . . . and virtually no prospect of understanding the observed phenomena. As we shall see in later chapters, ancient Chinese records attest to observed meteorite falls in terse, unembellished prose. Egyptian and Greek records from the dawn of recorded history recount stories of fireballs and meteorite falls. Some illiterate peoples clearly believed that meteorites were celestial objects (or beings) that fell from heaven, as we deduce from the documented Amerind practice of burying meteorites ceremonially, attended by beads and wrapped in blankets, in stone-lined graves. In many cultures, iron meteorites were assiduously sought for use by blacksmiths to make ceremonial objects and weapons. In some cultures, stony meteorites were venerated, or even, as in China until the twentieth century, ground up and eaten.

But Western culture was very reluctant to authenticate reports of meteorite falls. At first, Aristotle's doctrine of the geometrical perfection and spiritual (and hence immaterial) nature of the heavens stood as a barrier. It was simply not admissible that something as blatantly solid as a rock could have come from the heavens. The Church-sponsored Aristotelian doctrine of the perfection of the heavens also led seventeenth-century theologians to deny, in contradiction to Galileo, that there were craters on the Moon (the Moon is a celestial body and hence must be a perfect sphere) or that there were satellites in orbit around Jupiter (they are not counted among the seven heavenly bodies mentioned in the Bible, and therefore must not exist). This perfect uncoupling of heaven from Earth, of spiritual from material, persisted as a kind of living philosophical fossil for the next two centuries, predisposing authorities of the most diverse beliefs to ignore reports of meteorites falling from the sky

Then the new Establishment, the rising Protestant sects of northern Europe, secured its day in the sun of uncontested authority. Elated by their first opportunity to serve as Guardians of Truth and Traditional Wisdom, they weighed in with equally reactionary vigor.

Consider Martin Luther's assessment of Copernicus: "Men give ear to an upstart astronomer who tries to show that the Earth revolves, not the Sun and the Moon. This fool wishes to reverse the entire science of astronomy."

It is not that Western tradition has been wholly free of references to celestial phenomena. Biblical writings, which lie at the root of Western culture, make numerous mention of portents in the heavens. Perhaps the most spectacular and pervasive allusions to flying mountains, fireballs, and apparent impact phenomena are found in the book of Revelation, written around A.D. 50. Whether derived from traditional accounts or from visions, the description of future events in Revelation leans heavily upon the phenomenology of violent cosmic events. The sixth chapter of Revelation begins with the opening of the first of the six seals on the book of judgment, accompanied by "the noise of thunder." At the opening of each of the first four seals a horse goes forth (white, red, black, and "pale," respectively), reminiscent of many classical descriptions of comets as "stars with a mane like a horse," and usually explicitly connected to some form of death. But events accelerate with the opening of the sixth seal, when "lo, there was a great earthquake; and the sun became black as sackcloth of hair, and the moon became as blood; and the stars of heaven fell unto the earth, even as a fig tree casteth her untimely figs, when she is shaken of a mighty wind. And the heavens departed as a scroll when it is rolled together; and every mountain and island were moved out of their places."

We are given to understand in Revelation 7:3 that the serious work of destruction had not yet begun: most of the events so far described are associated with the heavens, not the surface of the earth. The correlation becomes perfect if we think of the "earthquake" as being an atmospheric, not a geological, phenomenon.

Chapter 8 of Revelation deals with the opening of the final, seventh, seal. Seven angels stand before God with trumpets, each marking a stage in the final drama. When the first angel sounds his trumpet,

> there followed hail and fire mingled with blood, and they were cast upon the earth; and the third part of trees was burnt up, and all green grass was burnt up.
>
> And the second angel sounded, and as it were a great mountain burning with fire was cast into the sea: and the third part of the sea became blood; And the third part of the creatures which were in the sea, and had life, died; and the third part of the ships were destroyed.
>
> And the third angel sounded, and there fell a great star from heaven, burning as it were a lamp, and it fell upon the third part of the rivers, and upon the fountains of waters;
>
> And the name of the star is called Wormwood: and the third part of the waters became wormwood; and many men died of the waters, because they were made bitter.

The fourth angel sounds, and the Sun, Moon, and stars are smitten and the sky of Earth is darkened. Wormwood here alludes to artemisia, an herb well known in biblical times and cited elsewhere in the Bible, noted for its extreme bitterness. But note that these waters are not merely unpleasantly bitter like the herb; they are lethally toxic. Why should a "falling star" (meteor) give the waters a bitter taste and make them deadly? The fifth angel sounds, and John "saw a star fall from heaven unto the earth: and to him was given the key of the bottomless pit. ... And he opened the bottomless pit; and there arose a smoke out of the pit, as the smoke of a great furnace; and the sun and the air were darkened by reason of the smoke of the pit."

Why should a star, falling to Earth, open a great pit and fill the air with enough smoke to darken the Sun? The central theme is clear and unambiguous: the events described in Revelation are of astronomical origin, and describe real physical events, not mere portents or symbols. Did John somehow know more about impact phenomena than any scientist before the present decade? Or, if he was simply borrowing a metaphor to convey his message, from what earlier source did he derive his inspiration?

We do know that stories of ancient meteorite, fireball, and impact events existed then, although we have no evidence that John had access to them. For example, Pliny the Elder wrote around A.D. 70 in volume 2 of his great compendium *Historia Naturalis* of the fall of a meteorite at Aegospotami (literally, "Goat River") in the Hellespont in 464 B.C., apparently near the very spot of the famous victory of the Spartan navarch (naval commander) Lysander (in league with the Persian satrap Cyrus) over the Athenian navy in the summer of 405 B.C. The account describes a very large meteorite the size of "two millstones," too large to be readily moved. The famous German naturalist Alexander von Humboldt commented in 1848 that this meteorite (if such it was) had long been lost, but that some hope still remained of finding it. It remains unfound to this day.

Pliny also asserted that the mathematician and astronomer Anaxagoras of Clazomenae had predicted the Aegospotami meteorite fall. Anaxagoras was a prominent participant in the political life of Athens. The great Athenian statesman Pericles had seized upon Anaxagoras' "clockwork universe" theory in much the same way that eighteenth-century European political theorists seized upon Newtonian mechanics as a political paradigm that freed them from theological bondage, allowing them to deal pragmatically with political matters. (In this sense, Pericles appears as an ancient counterpart of Machiavelli!)

The denouement of Anaxagoras's story should serve as a sobering warning to scientists who dare to meddle in public affairs: as part of a political campaign to discredit Pericles, Anaxagoras was prosecuted and fined for holding "impious doctrines," and left Athens to continue his studies elsewhere (in much the same way that professors of "Jewish physics" were driven from Nazi Germany by Hitler). A friend of both Anaxagoras and Pericles, the famous sculptor Phidias, was accused of embezzlement of state monies dedicated to the beautification of the acropolis, and Pericles' mistress Aspasia was also prosecuted. At about the same time, his two legitimate sons died of the

. . . .

plague. His heart broken, Pericles died soon after (429 B.C.). Evidently Anaxagoras's mathematical prowess, reputedly sufficient to predict the fall of a meteorite (a feat impossible today, and probably literally of mythic proportions), afforded little protection against political and religious enemies. But the *origin* of the story about Anaxagoras's predictive powers remains unexplained. Did he in fact predict a meteorite fall? His "clockwork universe" theories would not enable him to predict such an event because he could not possibly have known the orbit of a meteorite-sized rock before it actually fell. But he might very easily have predicted brilliant fireballs and aerial explosions around a certain date—such events often cluster around certain dates, when Earth passes through the intersection of an orbit followed by swarms of comet debris. Such events could have been predicted by anyone of Anaxagoras's time who collected and compared records of past fireballs.

Whatever his personal expectations, Pliny was a compulsive collector of data. His accounts of "meteors" (which until recently meant all atmospheric phenomena) are rich in strange and often incredible tales. He provides straightforward accounts not only of the fall of the Aegospotami meteorite, but of many other extraordinary events. He describes small meteoric lights that course swiftly across the sky, making long trails, with their front ends glowing, and of bolides (from the Greek βολιδας, meaning "javelins"), which make longer, more persistent, glowing tracks. He then talks of "blood and fire" falling from the skies, then of falls of milk, blood, and flesh. He tells of a fall of iron from the sky in Lucania in 54 B.C., and then of a rain of "wool"; of clanging armor and sounding trumpets in the sky, and of the apparition of an apparent giant comet to the Ethiopians and Egyptians. His accounts thoroughly confuse fact and fancy, observation and interpretation.

In the twentieth century such strange reports are linked firmly with the remarkable writings of Charles Fort. For decades, Fort combed the resources of the British Museum and the New York Public Library, collecting every observation that seemed to lie beyond the pale of contemporary science. He compiled pages of reports of falls of "blood" (red rain), black rain, and black dust; pages on falls of gelatinous goo; pages on mysterious sounds of distant gunfire coming from empty oceans; more pages on falls of frogs and fishes (though he cites no actual reports of rains of cats and dogs). His avowed purpose was to tweak the noses of the respectable scientific establishment. But, endowed with a puckish sense of humor, a breathless writing style, and unusual intelligence, Fort did not content himself with presenting a mere compilation of grotesqueries: he offered tongue-in-cheek "explanations" in the most extravagant possible terms, many of which make truly hilarious reading. So factually circumstantial and conceptually outrageous were his books (*Lo; New Lands; The Book of the Damned; Wild Talents*) that they have captured the imagination of at least two generations of readers. A measure of his impact is that Fort also contributed his name to the language: the word *fortean* can now be found in wide use with the approximate meaning "bizarre but possibly true."

Many traditionally fortean phenomena have since come to roost in the world of science. Falls of red dust and red rain are now well-documented phenomena associated with sandstorms in the Sahara Desert: even falls of red dust and rain in Barbados can now be traced back to Saharan dust by means of weather satellite imagery and precise chemical analyses. Most reports of falls of soot and black rain are intimately intertwined with the rise of the Industrial Revolution in England. Falls of frogs and fish have been linked to waterspouts and other rare but authentic meteorological phenomena. The "Barisal guns," so named after the frequent reports of distant gunfire heard in Barisal, Bangladesh, coming from the Bay of Bengal, are now understood in terms of a waveguide-like behavior of the atmosphere, channeling sounds over vast distances under thermal inversion layers. Inversion layers, in which the temperature increases with altitude and vertical mixing of the atmosphere is hindered, are common over the oceans and over certain cities such as Los Angeles, where they trap pollution close to the ground. The same process that makes aerial explosions audible over vast distances is also responsible for the long-distance communications prowess of whales, which are able to "sing" over vast distances by diving beneath the solar-heated thermal inversion layer in the ocean.

Pliny's catalog of fortean phenomena, however, drawing as it did upon mythic as well as documentary sources and presented without much semblance of critical thought, seemed better suited to medieval times than to the Age of Reason. Many of the stories related in his work fell, to the medieval mind, into the category of "celestial signs and portents" rather than "physical phenomena." Reliable eyewitness reports of actual events, when embedded in such a list, suffered a serious erosion of credibility. And of course, when events are so remote in time as to have taken on mythical dimensions, it is very difficult to extricate the original happening from the later fabrications and interpretations. Nonetheless, it is worth remembering that the people of ancient times, although equipped with a physical understanding of the universe that was vastly inferior to ours, were just as intelligent as we are today. They commonly observed the skies with eyes unobstructed by smoke, smog, and light pollution, and with minds untutored in theoretical preconceptions. Those who were out at night, mostly shepherds, were intimately familiar with the night sky in a way that modern city-dwellers cannot begin to imagine, and watched it in a darkness more true than most moderns have ever experienced. It was "shepherds in the fields by night," not the priests of Jerusalem, who saw the portents of the birth of Christ. Whatever unusual event occurred in the heavens, they noted it; and, though they lacked both a modern technical vocabulary and a theoretical understanding of physics, they tried to tell us what they saw.

From classical Greek times through the Middle Ages, the personification of celestial phenomena was widespread. It was traditional for princelings to commission elaborate genealogies alleging their descent from mythical heroes or gods. In ancient times, descent was claimed from (usually illegitimate) offspring from the affairs of the immortal gods with mortals; the inclusion of many prominent names from history and mythology in the family tree was deemed desirable. The Egyptians traced the origin of their civilization to Pythom, which was apparently a comet. In medieval times, members of the nobility and royalty traced their descent, by the most tenuous of reasoning and quite unburdened by evidence, back to Old Testament patriarchs, and thence onward, invincibly, to Adam. Many of the heroes and gods of these tabloid genealogies were in turn apparent personifications of astronomical bodies and phenomena observed anciently. The concept of the divine right of kings first arose millennia ago from the assertion that the royal families were literal descendants of gods (Venus, Mars, Jupiter, Saturn, Uranus, etc.) that we would today describe as celestial bodies.

Events in the sky have always been incompletely observed and reported. Even well within historical times, it has been the misfortune of most meteorites to fall under circumstances in which no certified expert witnesses, whether theologians or natural philosophers, were present to observe the event. Quite aside from the natural tendency of meteorites to fall randomly to Earth (and hence to sink without a trace into the oceans 72 percent of the time), those that fell on land more than a few centuries ago were most likely to fall in uninhabited areas. Those that fell near civilization generally fell in rural areas (very little of Earth's surface was urbanized, let alone suburban-ized, five hundred or one thousand years ago!), where they were either not observed or were observed and recovered by illiterate, uneducated peasants. It was quite unthinkable that the sophisticated urban theologian of A.D. 1000, raised in the stratospheric refinement of church and court, or the priest-professor of natural philosophy of the medieval universities, should take seriously the allegations of such rabble that stones fell from the sky. How could the testimony of an unlettered rustic stand against the authority of an Aristotle? How could a member of the most elite class of society deign to be in-structed by a man of crude speech and dress, wholly devoid of creden-tials? Why should any record be kept by (literate, educated) chroniclers of such fantastic allegations by (illiterate, uneducated) peasants? It is an astonishing but well-established fact that modern astronomers seeking accounts of ancient astronomical events (eclipses, comet apparitions, great meteor displays, meteorite falls, supernova explosions, etc.) find the records of medieval Europe to be sparse at best. Events, when reported by European sources, are often treated as heavenly or prophetic portents of the ever-impending millennium. The true treasure troves of such informa-tion are Chinese, Arabic, Babylonian, Persian, and Korean records, all of them free of the influence of Christian chiliastic (millenni-alist) expectations.

But the intellectual climate of Europe began, quite without intent or planning, to change. The rediscovery of the writings of classical Greek, Roman, and Egyptian sages by Crusaders in the Holy Lands and in Moorish Iberia brought with it a reawakening of interest in the past. The translation of Aristotle into European languages un-veiled an enormously richer knowledge of the natural world than was previously available in the West, a knowledge drawing upon a wide range of observations of nature. Aristotle unquestionably spoke with more authority about the world than any writer in thirteenth-century Europe. His writings were revered by some, but suspected of pagan influence by others, and helped pave the way for the rise of the universities. The universities founded in the great cities of Western Europe in the thirteenth and fourteenth centuries, at Rome and Padua, Paris, Oxford, Cambridge, and many others, whether founded by civil or religious authorities, were brought under control of the Church. The purpose of these universities was not to discover new knowledge, but to perpetuate existing doctrine, especially in the three established preprofessional areas of theology, law, and medicine. Undergraduate education as we normally think of it did not exist. There were no liberal arts, no research, no experimental sciences. All doctrine was screened for its compatibility with accepted theological views. The natural philosophy of Aristotle, the Earth-centered cosmology of Ptolemy, and the physiology of Galen were taught as perfected knowledge, as reliable dogma, rather than as illustrations of the importance of observation and experiment. But the universities attracted the learned and the curious: dogma collided with curiosity. Dogmas that could not withstand scrutiny withered, while those that accommodated observations and questioning prospered. And always, when doctrines were questioned, it was the Church, wearing the ill-fitting cloak of Aristotelian certainty, that defended the status quo.

The six great waves of the Black Death, which shattered the institutions of Europe between 1347 and 1400, ushered in the age of systematic geographical exploration. The European age of exploration began in the early 1400s at the impetus of a single man, Prince Henry the Navigator of Portugal, extending first to the Azores and Canary Islands and the coast of West Africa. By the end of the century much of the globe had been redrawn. North and South America had been added, and much older geographical lore had been discredited. Influenced by these discoveries, starting in the late 1400s, thinkers in many disciplines began to apply new standards of objective inquiry to their areas of interest. Many of these thinkers saw their role as one of declaring the independence of their discipline from theological domination. With this attitude, which was opposed at every turn by powerful elements in the Church, it was inevitable that a deep schism should open between the worlds of faith and intellect. Earth and Heaven were divorced.

The vaunted independence of Earth from cosmic influences, conceived at first as a demythologization of science and a liberation of geology from medieval influences, could only be maintained by ignoring certain types of evidence long in hand. For example, just weeks after Columbus reached the New World, a spectacular meteorite fell in Germany. On November 7, 1492 (Julian calendar), shortly before noon, a brilliant fireball crossed the sky over Basel heading northward. A young artist, Albrecht Dürer, saw the spectacular sight, and painted the scene a little over a year later. Loud explosions were heard as far away as Lucerne and the canton of Uri. The shock wave from the explosion weakened to the north as the fireball decelerated. Just outside the Alsatian town of Ensisheim a boy heard the blast, looked up to see a cloud of dust and debris, and was astonished to see a large rock fall from the sky and bury itself about a meter deep in a field nearby. A crowd quickly gathered and dug out the stone, which at that time weighed close to 140 kilograms (300 pounds). These events by themselves would probably have sufficed to guarantee the survival of the story down to our time. But, in a strange stroke of luck, this fall occurred as Maximilian and his armies were approaching Ensisheim. Maximilian, then sporting the title "Romischen kuning" (Roman King) rode into town on November 26 and announced that the fall was a sign from God. He ordered the stone displayed in the Ensisheim church, then went forth to win a battle with the French, thus cementing the reputation of the stone as a divine harbinger.

The Swiss poet Sebastian Brandt, who chanced to observe the fireball from Basel, penned a commemorative verse that was quoted in part at the beginning of this chapter. This verse appears, with considerable spelling variation, in several broadsheets and posters published in late 1492 and early 1493, all of which emphasize the miraculous and prophetic aspects of the fall. The then recent invention of the printing press made it possible to disseminate the news of the fall unusually quickly and widely, with unprecedented impact on public opinion. Ursula Marvin, of the Harvard-Smithsonian Center for Astrophysics, has turned up some chronologies of the century that list the Ensisheim fall as the only event recorded for the year 1492!

But at the time this event occurred, the gulf between intellect and faith was gaping ever wider. In 1513, Machiavelli (*The Prince*) developed a theory and practice of politics that had no basis in theology. Martin Luther posted his ninety-five Theses in 1517, positing a view of religion in which, as some wise historian once phrased it, the *corpus* of the Church would not smother the *Christus*. Erasmus (1520) developed the central ideas of humanism without deference to Christian dogma. Copernicus (*De Revolutionibus*, 1543) demythologized astronomy by challenging the traditional Ptolemaic doctrine of the centrality of the Earth in the universe. Tycho Brahe (in 1562, at the tender age of sixteen) further showed by direct observation that the widely respected Alphonsine Tables, showing the positions and motions of heavenly bodies, were not even close to correct. Vesalius (1543) demythologized the medieval standard of knowledge of the human body, showing the error of many of Galen's doctrines of human physiology by recourse to the study of cadavers. The spirit of the age was to compare theory with observation; to establish experimentation and observation, not tradition or ancient authority, as the final arbiter of truth. It was the dawn of the age of empiricism.

This spirit of questioning, innovation, and experimentation could not be excluded from the realm of practical politics forever. During the time of Cromwell's Commonwealth (1649–1660) the democratic principle first, however feebly, raised its head in England. Progressive extensions of the right to vote to ever larger segments of the population accompanied the realization that the rights of royalty could be decided by the people, not vice versa. The growth of democracy in Europe was accompanied by general acceptance of a Newtonian view of the universe: that all physical objects, including "celestial" (astronomical) bodies as well as mechanical devices, people, and societies, move in accord with a set of universal laws. All reference to celestial portents and divine extralegal (catastrophic; spiritual) interference in earthly events was excised from the Newtonian universe. The religion of the time, though still unabashedly Christian (Newton spent his later years writing books on theology), took on aspects of deism. Benjamin Franklin, a progressive scientist who espoused Christian morality, favored a universe in which the Creator established the new creation and its laws, and left it to work out its own fate in accordance with those laws. Newton's physical laws, though incompletely known, were all that were needed to explain heavenly phenomena: in principle, even the bizarre and "unpredictable" motions of comets were explicable. The temper of the time was at first to deny that the prophetic visions and apparitions, portents, and divine interventions of ancient days any longer took place. It was only later that it became fashionable to deny that they *ever* took place.

By the end of the seventeenth century, Newton's laws, in the hands of Edmund Halley, had been found to account for the motions of comets, thus at last demythologizing the most intractable of the "celestial portents." Yet another Establishment arose, this time constituted of the practitioners and defenders of Newtonian physics. Newton himself, finding to his surprise that the orbits of comets could intersect Earth, hastened to deny that such collisions could have any negative effect, again affirming the lack of connection between heaven and Earth. Sir John Pringle, M.D., F.R.S., a contemporary of Franklin's, alleged, following Newton, that all heavenly bodies had been placed there by the Creator for purely beneficent purposes.

Thus the acceptance of many rare and spectacular celestial phenomena as physical rather than spiritual paradoxically went hand in hand with a desire to uncouple astronomical events from the conduct of terrestrial affairs. It is interesting that this movement, along with the founding of the French and British Royal Societies, occurred at the peak of a great wave of millennialist agitation among the Puritans and other Protestant sects. Millennialist expectation of the imminent Second Coming of Christ and the destruction of the Roman Catholic and Anglican churches provided much of the impetus for Cromwell's ascent to power. The Puritan worldview was exceptionally simple: the world consisted of a vast number of powerful secular establishments, all of which were doomed. The Puritans had no more interest in astronomy or physics than in the fine points of Catholic theology. But they were deeply interested in, and even obsessed by, portents. No less a Puritan divine than Increase Mather penned "An Essay for the Recording of Illustrious Providences," a plea for the systematic collection of the effects of various kinds of natural disasters, complete with judicious assessments of the spiritual state of each person who either fell victim to or miraculously escaped from these disasters. Clearly, Mather's purpose was not to understand lightning, storms, fireballs, meteorite falls, and the like—rather, he was interested in these events as morality plays.

For centuries the Ensisheim fall remained the only case in which eyewitness reports of a fall were combined with preserved meteoritic material. But other fireballs and meteorites were to come to the attention of geologists and chemists. A brilliant meteor, seen over most of Italy in 1676, generated hundreds of eyewitness reports that were collected and analyzed by Montanari, who published his work in a scholarly booklet, *Fiamma Volante* (Flying Flame). Montanari estimated the height of the fireball as about fifteen French leagues. When this conclusion was attacked by another author (Cavina), who claimed that the height was closer to forty-five leagues, Montanari was too much the gentleman to argue. But his assistant Guglielmini, described in the *Histoire de l'Académie des Sciences* (Paris) as a "truly zealous disciple of Montanari," joined the fray, widely publicizing this event and the potential for measuring fireball altitudes by geometrical analysis of widespread observations. Unfortunately, no meteorite was recovered from the Italian fireball, and no fireballs were seen in conjunction with several of the earliest discoveries of distinctive meteorites. A mass of unusual metal, completely unlike any known terrestrial rock, was found lying on the ground in Senegal in 1716. Then a very strange piece of iron from Siberia was studied by Peter Simon Pallas in 1771, and a number of large pieces of iron, found at Campo del Cielo, Argentina, were described by Don Rubin de Celis in 1783. Collectors and mineralogists traded samples back and forth, but consensus on their origin was lacking. Since none of these meteorites had been observed to fall, proof of their extraterrestrial origin was impossible. It became increasingly important that observed falls be linked to recovered samples of meteorites. But until that linkage could be established, the religious and mythic interpretation of the supposed prophetic meaning of fireballs became a great barrier to the general acceptance of the fall of meteorites from the sky.

In the wake of the Commonwealth, with the restoration of monarchy, the Royal Society of London was founded in 1665–1666. Some historians have argued that the Society was a welcome counterweight to the fiercely anticlerical and anti-intellectual climate fostered by the Puritans. The corresponding institution in France, the Société Royale, dates from the same year. Unlike his counterparts of earlier centuries, the gentleman-scientist of the seventeenth-century Royal Society found it possible to entertain the reports of peasants and other illiterate observers. The democratic principle found further expression in the American Revolution of 1775–1783, and finally in the French Revolution of 1789–1799. By 1800, democratic principles were firmly entrenched in western Europe and America. One might hope, and even expect, that conservative reaction against the reality of meteorites would vanish in this new climate. But the attitude of scientists, that the world is governed by comprehensible physical laws that can be discovered through observation, experimentation, and analysis, became simplified in the public mind: anything that was already understood in physical terms must be true, whereas anything that was not already understood in physical terms must be superstitious nonsense. Most damaging of all, some scientists fell into this very trap. Then as now, falsely equating the *understood* with the *true* creates a burdensome new body of dogma and installs a new caste of Guardians of Truth and Traditional Wisdom.

The effects of revolutionary political changes in America and France coincided with the resolution of a ferocious debate within geology. At the beginning of the revolutionary age it was commonly believed that the Earth was young, perhaps only a few thousand years old, an age compatible with the most extreme interpretations of scripture by Bishop Ussher of the Church of Ireland, who dated the creation as occurring on October 23, 4004 B.C., at 9:00 A.M. Ussher, acting on behalf of the Church of England's offshoot in Ireland, as yet another self-appointed Guardian of Truth, left a dismal inheritance to his successors. Given such a minuscule interval of time to hold the entire history of Earth, it was evidently necessary to attribute the enormous thicknesses of sedimentary deposits revealed by the studies of field geologists to a handful of very recent, extremely catastrophic events. The most notable of these was taken to be Noah's flood. The fossil record of earlier life on Earth, already becoming known from the study of thick fossil-bearing sedimentary deposits in Devon and Wales, had to be dismissed in some manner. Some asserted that the fossils were merely accidental structures of inorganic origin; some ascribed them to an attempt by the Devil to delude mankind, or by the Creator to test man's faith (a doctrine as antithetical to the teachings of Christian scripture as to geology itself). A more liberal establishment opinion held that the fossils were of true biological origin, but were remnants of "that accursed race which perished with the Flood."

In 1785 the Edinburgh gentleman-scientist James Hutton published his *Theory of the Earth with Proof and Illustrations,* which asserted for the first time the uniformitarian principle: that all the work of geology was achieved not by a few catastrophes, but by the familar slow processes of erosion, sedimentation, et cetera, that we witness in our daily lives, extended over many millions of years of time. Baron Georges Cuvier, the great French paleontologist who pioneered in the study of fossil fishes, mollusks, and mammals, laid the biological basis for just such an interpretation of geologic time. Charles Lyell, in his 1830 book *Principles of Geology,* further developed the uniformitarian view with such success that the opposing school of catastrophism ceased to have influence in mainstream geology. Thus the vast age of the Earth became the central tenet of geology. The idea that catastrophic events could occur on Earth fell into disrepute, along with the idea that cosmic (astronomical) events could have any effect on Earth. The fresh and revolutionary insights of Hutton, Cuvier, and Lyell, which were derived directly from observation, became as it were stuffed for museum display. Transformed from empirical evidence to dogma by succeeding generations of geologists, uniformitarian thought degenerated into yet another -ism. The latest crop of Guardians of Truth and Traditional Wisdom had found a new line to defend, and they dug in deep enough to hold their position for two centuries.

But the meteorites were a potential embarrassment to the purest forms of uniformitarianism. Fortunately, some scientists saw them as posing tractable scientific questions and offering new insights. In

. . . .

1757, reporting on a brilliant fireball accompanied by loud explosions, and again in 1759, reporting on two fires set by falling meteorites, the journal of the French Academy editorialized:

> *These meteors are not rare; one finds examples of them in our Memoires,*
> *in the philosophical Transactions [the Philosophical Transactions of the*
> *Royal Society of London], and elsewhere. The more these observations mul-*
> *tiply, the more one finds that, on the contrary, they are common; but what*
> *makes observation (of them) important is that, at the height at which they*
> *are elevated, they might perhaps one day inform us of the height of our*
> *atmosphere, or at least that it is much more extensive than has hitherto*
> *been supposed.*

Thus, stepwise, the study of meteors and meteorites moved ever closer to what could be recognized as science.

The secularization of meteorite studies followed the political winds of change. Wolfgang von Goethe, recently liberated from his university studies of law, visited Ensisheim in 1771 to see the purported meteorite. The logic by which he proceeded to ridicule this object seems to be that since the authenticity of the meteorite was attested to by the Church, it must therefore be a fraud! This scorn was echoed almost unanimously by the Guardians of Truth as late as 1791, when the editor of the practical science periodical *Journal des Sciences Utiles*, Pierre Bertholon, similarly mocked the unanimous reports of the residents of Barbotan, France, attesting to the fall of a shower of stones on July 24, 1790. Meanwhile, the Ensisheim stone, long protected by Maximilian's edict, resided in the parish church until removed by French revolutionaries to a museum in Colmar in 1793. This act of removing Europe's one witnessed meteorite from a church to a museum seems symbolic of the age.

Perhaps the first meteorite expert in the world was the German scientist Ernst F. F. Chladni, who diligently searched for reports of meteorite falls in old records. Chladni collected accounts of the falls of eighteen meteorites from Greek classical times up to contemporary events, including the Ensisheim stone, in a booklet published in 1794. Almost at once, the sky began to rain witnessed meteorites. On June 16 of that year a well-documented shower of stones fell near Siena in northern Italy, and a large stone fell at Wold Cottage in Yorkshire, England, on December 17, 1798. Then, in 1800 and 1801, a number of reputed meteorites were analyzed by E. C. Howard in England, who found striking chemical similarities between them, while finding no similarities to local rocks. The high nickel content of meteoritic iron emerged at about this time as an accepted indicator of nonterrestrial origin. Meteorites, wherever they came from, were not of Earth.

This watershed of opinion is well documented by returning to Goethe at the time of a visit to Göttingen in 1801, when he received an account of the 1798 meteorite fall in Benares, India, from Professor Johann Friedrich Blumenbach. Goethe's journal for June 7, 1801, records his acceptance of the reality and extraterrestrial origin of meteorites.

The transition of public opinion in France took slightly longer. In 1803 the French chemist Antoine-François de Fourcroy analyzed several of the same stones studied by Howard, and compared them to samples from the Ensisheim meteorite. He likewise became convinced of their authenticity. With a most peculiar kind of revolutionary fervor, he wrote, "I could distrust the imagination of a learned man, but I would place all my faith in the testimony of an ignorant person, because, by nature, the ignorant person has no imagination."

This backhanded encomium (praising with faint damns) can scarcely be regarded as a victory for equality or fraternity, let alone liberty, but does at least represent one small step of progress toward accepting eyewitness reports. De Fourcroy concluded that Ensisheim belonged to the class of fallen stones rather than natural (terrestrial) stones. But the strongest case for meteorite falls arose from the fall of a shower of some twenty-five hundred stones at L'Aigle, Normandy, on April 23, 1803. The famous French scientists Jean-Baptiste Biot, Pierre-Simon Laplace, and Siméon-Denis Poisson, investigating on behalf of the French Academy of Sciences, joined in accepting the legitimacy of this fall. The source of the stones that fell from space presented a puzzle: where could they possibly have come from? Biot, Laplace, and Poisson offered the opinion that meteorites were ejected from volcanoes on the Moon and fell from the Moon to Earth, striking the atmosphere at very high speeds.

As far back as 1665, the English physicist Robert Hooke had carried out numerous experiments on the formation of lunar craters by dropping bullets into clay and mud. He concluded that the craters made in this way were strikingly similar to those on the Moon, yet he declined to conclude that impacts caused the lunar craters because "it would be difficult to imagine whence these bodies should come." But once the fall of rocky bodies from space onto Earth was documented, a new insight into the origin of lunar craters became possible. But that meant that *very* large rocks were falling from somewhere *onto* the Moon ... and onto Earth as well! That story is told in chapter 2.

The American counterpart of the L'Aigle meteorite was the fall of a stone meteorite in Weston, Connecticut, on December 14, 1807. Professors Benjamin Silliman and J. L. Kingsley of Yale University investigated the fall and pronounced it credible. There is a long-standing but unverifiable tradition that the naturalist-president Thomas Jefferson, on reading their report, remarked that he "would find it easier to believe that two Yankee professors would lie, than that stones should fall from the sky." Faced with no clear explanation of *how* stones might fall from the sky, this paragon of democratic and scientific ideals evidently found it easier to deny that they *did* fall. Thus did Jefferson win his honorary membership in the Order of the Guardians of the Truth and Traditional Wisdom.

Although debate about the origins of meteors and meteorites was to continue for many years, it was generally accepted by 1820 that they were extraterrestrial material. A review written by the American physician William G. Reynolds in 1819 describes the four major theories of the era: that meteors are purely electrical phenomena; that they are (in Halley's opinion) natural nonterrestrial solids that Earth encounters on its annual trip around the Sun; that they are celestial bodies placed in the heavens for beneficent (if inscrutable) purposes by the Creator; and that they consist of highly inflammable materials of earthly origin that condense high in the atmosphere. Reynolds, voting late and choosing poorly, decided in favor of the last of these. But within a few years three of these principal ideas of earlier times were generally abandoned. It became generally accepted that they were solid natural Sun-orbiting bodies that strike Earth's upper atmosphere at high speeds. The study of meteorites could now be pursued with the intention of finding out the nature of their "parent body," the planet, moon, or asteroid from which they came.

With over ten thousand authenticated meteorites in our collections, modern scientists have a good idea of the nature of the solid materials that fall to the ground from space. These materials fall into three general classes. The first of these, the irons, are composed of a natural stainless steel, an alloy of iron, nickel, cobalt, and precious metals, with perhaps 1 percent of impurities such as carbon and sulfide minerals. The second class, the stones, are made up mostly of fairly familiar silicate minerals, but they usually contain metal particles similar in composition to iron meteorites. Third are the stony-irons, which contain about 50 percent each of metal and silicates. These three major classes can in turn be divided into some fifty distinct groups on the basis of their chemical composition and mineral content. The meteorites show a very wide range of strength, from enormously strong and tough metal that can pass through Earth's atmosphere almost without disruption to soft, easily crushed sediments that rarely survive their blazing passage through the atmosphere.

Five out of every six known meteorites are very ancient and moderately strong stones, called ordinary chondrites, that contain from a few percent to a maximum of about 30 percent of metal. Most of these ordinary chondrites belong to two very similar major groups which differ principally in their iron content (the high-iron or H chondrites and the low-iron or L chondrites). Some 2 percent of all meteorites are the even stronger stony achondrites. The achondrites, which are not numerous, are clearly the result of melting of chondrites, followed by separation of its components under the influence of gravity. During such a separation (or "differentiation"), the densest materials, metals and sulfides, sink, and the least dense materials, the silicates, rise to the top to make the achondrites, much as the rising slag from the melting of the early Earth formed the crust while the metals formed the core. The complementary dense, strong core-forming material separated during the formation of the achondrites makes up the 3 percent of all known meteorite falls that we call irons. About 1 percent of all known meteorites are samples of the contact between achondrites and irons, the stony irons, all of which are also extremely strong.

About 1 percent of all recovered meteorites are the very peculiar carbonaceous chondrite stones. These very weak stones are rich in water, which is bound up in both hydrated salts and clay minerals. They also, astonishingly, contain abundant organic material. We know they are meteorites because they are observed to fall and because they all have been directly exposed to cosmic rays in space. Cosmic rays, the highly energetic protons and other nuclear particles that pervade the galaxy, break apart the nuclei of atoms in their targets and produce distinctive signatures in the form of short-lived radioactive nuclei which can be detected in freshly fallen meteorites. Some of the carbonaceous (C-type) chondrites are physically quite weak. Of them, the Ivuna-type (CI) chondrites are the ones with the highest amounts of water (up to 20 percent) and organic matter (up to 6 percent). They are about as strong as dried clods of mud. Indeed, as we discuss in chapter 4, C chondrites are so fragile that hardly any of their material survives passage through the atmosphere. The drier and less carbonaceous Murchison-type (CM) chondrites are also quite weak, although they are about ten times as strong as the CI chondrites.

All meteorites contain higher concentrations of sulfur than typical terrestrial rocks, and give off a strong odor of sulfur compounds upon heating. "Brimstone," along with fire the classical harbinger of diabolical manifestations, is merely an archaic term for sulfur. Since volcanic activity was well known in the classical world, it comes as no surprise to us to find fire and brimstone mentioned in connection with a subterranean hell. But to find them also mentioned in connection with aerial malign phenomena actually makes perfect sense if the events described are connected with meteorite falls.

The very different strengths of the classes of meteorites favor the preferential survival of hard stones during entry into the atmosphere and impact on Earth. Irons often reach the ground as a single large body: iron meteorites massing up to thirty metric tons have been documented. The very weak carbonaceous meteorites often fall as showers of tiny fragments with masses of grams. Thus crushing strength provides a selection effect that discriminates powerfully against the survival and retrieval of carbonaceous chondrites, and in favor of irons and strong stones. It is very likely that irons are rarer, and carbonaceous material much more common, in the asteroidal and cometary debris that strikes the top of Earth's atmosphere than the statistics of recovered meteorites would suggest. It is very important whether the bullet that hits your planet is made of steel or dust!

The total rate of recovery of freshly fallen meteorites is nearly the same in all densely populated areas: about one recovered meteorite per year per million square kilometers of ground area. These rates are achieved in areas such as the states of New Jersey and Connecticut, the Netherlands, Japan, and certain parts of India and China. In less densely populated areas the recovery rate is of course correspondingly lower.

Simultaneous to, and quite unconnected with, the debate over the origin and authenticity of meteorites, astronomers in the early 1800s were reporting the discovery of a whole new class of solar system bodies, one that would, eventually, hold the key to the understanding of meteorites. On the first night of the nineteenth century, January 1, 1801, Giuseppi Piazzi of Palermo, Sicily, discovered the asteroid Ceres. On March 28, 1802, Heinrich W. Olbers of Bremen found the asteroid Pallas, and the unveiling of the richness of the asteroid belt was begun. But it was not until much later that any direct connection between asteroids and Earth could be established. We shall return to this story in chapter 6.

2

TARGET: EARTH

The author feels that he can announce the following facts as absolutely proved:

First: That at this locality (Meteor Crater, Arizona) there is a great hole or crater in the earth which corresponds in all respects, except in its gigantic scale, with impact craters formed in rock by projectiles of considerable size moving at considerable velocities.

Second: That in and around this hole and below its bottom to a distance of over 1,400 feet below the present surface of the plain surrounding it, and the original surface of the place where this hole was formed, every indication of either volcanic or hot spring action is positively absent.

Third: That in and about this hole all signs which might be expected of the impact of such a great projectile are present.

Fourth: That upon the surface of the rim and upon the surrounding plain there has been found and still exists a large quantity of meteoric material, and that the distribution of this material is symmetrical with a line passing through the center of this hole.

Fifth: That this meteoric material was deposited at the same instant of time as that at which the hole was made.

Sixth: That in and around this hole is an enormous quantity of pulverized rock, produced from the strata penetrated by the hole, in a state of subdivision which can be produced by a violent blow, but cannot be produced by forms of natural erosion. . . .

In view of these positively established facts, the author feels that he is justified, under due reserve as to subsequently developed facts, in announcing that the formation at this locality is due to the impact of a meteor of enormous and hitherto unprecedented size.

B. C. TILGHMAN, 1905

Scientific acceptance of the fact that large bodies from space can and do impact Earth did not come easily. The story of the authentication of the first impact crater serves as an excellent illustration of how impact craters are authenticated, how impacts can affect Earth's surface—and how defenders of entrenched theories protect their turf.

* * *

A rumor was abroad in northern Arizona in the late 1870s that chunks of silver had been found lying on the ground in a depressingly arid and barren stretch of desert east of Flagstaff, near the peculiar circular hill known as Coon Butte. In March of 1891 a sample of this metal was sent to an assayer in Denver with the message that there were carloads of the stuff lying around. The assayer responded with a report that the metal was mostly iron, but contained 1.8% lead and 0.025% silver, with a trace of gold. Unfortunately, this analysis was grossly in error: we now know that the 1891 sample contained negligible amounts of lead, silver, and gold, but did contain 7.9% nickel, an amount very rare in nineteenth-century commercial alloys, but (as noted in chapter 1) very representative of metallic meteorites.

Fortunately, another sample was sent to the Atlantic and Pacific Railroad, which forwarded it to Dr. A. E. Foote, a mineral dealer in Philadelphia. Dr. Foote was sufficiently astonished by the sample to undertake an immediate trip to Arizona, arriving in June 1891 with a crew of five assistants to recover what he confidently expected to be samples of iron meteorites. Foot described Coon Butte as a nearly circular elevation some twelve hundred meters (three-quarters of a mile) in diameter, surrounding a deep, precipitous basin. The flat-lying sediment layers in the country rock had been lifted and tilted outward all around the crater rim. He was unable to discover any trace of volcanic activity at the site, but he did recover 137 iron meteorites from the vicinity of the rim. The largest weighed in at 154 and 201 pounds (70 and 91 kilograms). His cautious written report emphasized the nonvolcanic nature of the crater and the abundance of irons, and mentioned the presence of tiny black diamonds in the iron, but did not mention a possible meteoritic impact origin for the crater. Nonetheless, he verbally expressed to a colleague, Charles R. Toothacker, that the crater had clearly been produced by a huge meteorite.

One of the great American geologists of the time, Dr. Grove Karl Gilbert of the U.S. Geological Survey, read Dr. Foote's published account in the *Proceedings of the American Asssociation for the Advancement of Science* later that year. Gilbert was at that time involved in a study of the Moon, an activity that drew the ire of a congressman who gruffly complained that, "so useless has the Survey become that one of its most distinguished members has no better way to employ his time than to sit up all night gaping at the Moon." This research was to culminate in Gilbert's classic 1893 paper demonstrating that the craters on the Moon are caused by impacts, not by volcanism. It is perhaps the most quoted paper published by any USGS employee of the century, and the one that has best maintained current interest to this day. Thus Gilbert, upon reading Foote's article, was intrigued by the possibility that cratering on the Moon and on Earth might have a common cause. He resolved upon a more thorough investigation of the crater by a trained field geologist.

Gilbert promptly arranged for a USGS colleague, W. D. Johnson, to pay a visit to Coon Butte later that year. Johnson, accustomed to thinking of meteorites as objects that could be held in one's hand, and craters as products of volcanism, reached the impeccably conservative conclusion that Coon Butte was only a volcanic feature that happened to have been formed by nonobvious processes. Gilbert, steeped in his observations of the Moon, was not convinced by Johnson's assessment. He and Marcus Baker, of the U.S. Coast and Geodetic Survey, followed up with a visit to Coon Butte in November of the same year. Two pieces of evidence from this visit persuaded Gilbert that there was nothing to the impact hypothesis. First, a magnetic survey of the vicinity of the crater turned up no evidence of the presence of a massive body of iron beneath the crater floor. Second, his calculations indicated that the volume of rim material was very nearly equal to the volume of the central basin, leaving no room for the presence of a great mass of meteoritic material. Evidently, Gilbert imagined crater formation as akin to dropping a heavy rock into wet sand from a height of a few feet. Indeed, Gilbert, retracing the path followed by Hooke in 1665, performed a series of impact experiments in his laboratory at Columbia University in New York. Limited to available technology (dropping marbles into bowls of oatmeal, etc.), Gilbert had difficulty understanding why lunar craters (and Meteor Crater) are so circular, since his experiments frequently yielded very elongated craters from low-velocity impacts at low grazing angles. He also noted that the presence of the marble displaced a large volume of oatmeal, resulting in a crater that could be completely filled by only a fraction of the rim material. And of course the marble was there to be dug up.

In 1892 Gilbert presented his conclusions to the Philosophical Society of Washington, of which he was the outgoing president: craters on the Moon are made by impacts. When Gilbert's final conclusions were published in 1896, he favored the volcanic hypothesis for Coon Butte, but still found the presence of irons to be a serious loophole in his argument. His conclusions, however cautious and lukewarm, were accepted as definitive by the world. Coon Butte was volcanic in origin; but the volcanic process had inconveniently left absolutely no trace of itself. A word was coined to describe such features, which "everyone knew" were volcanic, but contained no evidence of volcanic processes: *cryptovolcanic*, in which the "crypto" cryptically denotes "hidden."

In 1902 D. M. Barringer, a mining engineer, and his physicist friend, Benjamin C. Tilghman, heard of the crater. Barringer was an expert in ores and mining techniques; Tilghman on explosions. Examination of small samples of iron from the crater convinced both of them that the crater was an impact feature made by a huge projectile of iron-nickel alloy—and thus that there was a great fortune to be made digging it up. Over the following few years the two carried out at least ten expeditions to the crater, studying the rim deposits, making magnetometer surveys, drilling the crater floor, and searching for irons. In 1905, Barringer and Tilghman published accounts of their research for their respective communities, concluding that about 10 million metric tons of the meteorite responsible for the crater were still buried there. Henry Norris Russell, dean of American astronomers and chairman of the astronomy department at Princeton University, was convinced that this was indeed an impact crater. His colleague W. F. Magie from Princeton did a magnetic survey of the crater and agreed that it was of impact origin, and that much of the meteor remained buried in it. But despite the new evidence, the general weight of geological opinion continued to follow Gilbert's lead.

Fortunately for posterity, Barringer was a stubborn man. He and Tilghman continued to pursue the giant mass of metal that Barringer still erroneously believed to be present under the crater floor. Drilling uncovered small particles of metal, vast quantities of silica dust, and fused quartz glass of the kind presently called an impactite. They recognized that temperatures high enough to melt quartz had never been achieved in any volcanic eruption ever recorded, however violent and extensive.

In 1923, when a drill jammed at a depth of 417 meters, Barringer believed he had finally reached the main mass of iron, but further excavation in the following years failed to substantiate that conclusion. In 1929 Professor Forest Ray Moulton, an eminent theoretical physicist and astronomer from the University of Chicago, finally convinced Barringer, only months before his death, that the violence of the impact that created the crater would assure virtually complete

· ·

vaporization of the projectile, which would in any event not have massed ten million metric tons but only about one hundred thousand metric tons at the time of impact. Such a small mass is implied by properly accounting for the enormous kinetic energy of projectiles traveling at speeds in excess of 10 kilometers per second (6 miles per second; 21,600 miles per hour). Thus, thirty-eight years after the beginning of scientific exploration of the crater, there was at last the outline of a consistent theory for its origin—and that theory had "impact" written all over it.

Still, general acceptance was slow in coming. The famous 1911 edition of *Encyclopaedia Britannica* utterly ignored Meteor Crater and the associated Canyon Diablo iron meteorites. Its map of Arizona, recently updated because of Arizona's pending admission to statehood (in 1912), actually *omits the crater*. The article on the Moon characterizes lunar craters as "remarkable" but offers no opinons on their origin. We may suppose that their rejection of an impact origin for Meteor Crater relied upon Gilbert's authority. But then the omission of mention of an impact origin of lunar craters would have to be predicated upon a paradoxical rejection of Gilbert's authority. Clearly the editors felt uncomfortable with discussions of origins. We can perhaps detect here a lingering trace of Darwin's dictum that ultimate origins are unknowable and, in any event, outside the scope of science. Or perhaps they felt an irresistible urge to act as Guardians of Truth.

Moulton's introductory textbook *Astronomy*, published in 1931, contains a clear exposition of explosive cratering. The 1947 edition of the classic textbook *Astronomy* by Henry Norris Russell, Raymond S. Dugan, and John Quincy Stewart, their first revision since 1926, clearly describes Meteor Crater as a probable impact feature, and emphasizes that there is an abundance of meteoritic iron, but no evidence of volcanic activity within miles.

One reason for the reticence of the educated to accept impact cratering on Earth is that the debate on the origin of nonvolcanic craters was written in the shadow of the greatest conflict in the history of geology, the uniformitarian-catastrophist debate. As was discussed in chapter 1, unanimity had finally been achieved in the nineteenth-century geology community that the uniformitarian principle was valid: catastrophes were not only unimportant factors in geological change, they were nonexistent. In the context of such an attitude, Meteor Crater was a serious embarrassment, a mile-wide hole in their most fundamental theory.

. . . .

Once the first impact crater had been authenticated, other candidates quickly became known. In 1926 a geologist, A. B. Bibbins, published a note describing some heavily weathered craters with associated badly oxidized iron, that he had seen near Odessa, Texas, several years earlier. D. M. Barringer Jr. investigated the crater and found it strangely reminiscent of Meteor Crater. In 1927 a cluster of craters on the island of Saaremaa, Estonia, was described in a paper. And in 1927 the news reached the West of an expedition led by Leonid Kulik into the heart of Siberia in search of a giant meteorite crater that was reportedly formed in 1908. We shall have much more to say about that search in the next chapter.

An investigation by geologist Malcolm Maclaren of Lake Bosumtwi, in what is now Ghana, disclosed a crater some 10 kilometers in diameter with a 200–275-meter-high rim. Since no volcanic materials or processes were in evidence, and no native iron was found, Maclaren concluded that it was probably an impact crater produced by a stony meteorite.

In 1932 an explorer, with the impeccably British name of H. St. John B. Philby, set forth on an expedition to discover the whereabouts of the ancient city of Wabar in the Rub' al Khali, the vast central desert of the Arabian Peninsula. That city had, according to legend, been destroyed in ancient times by fire from heaven. According to that tradition, the site of the ruins was marked by a great lump of iron the size of a camel. The destruction story is reminiscent of the scene described in chapter 19 of Genesis: "Then the Lord rained upon Sodom and Gomorrah brimstone and fire from the Lord out of heaven; . . . And he overthrew those cities, and all the plain, and all the inhabitants of the cities, and that which grew upon the ground." The traditional date of this event is about 1900 B.C., which coincides rather closely with the rise of Babylonian astrology as an attempt to interpret celestial portents (not cast personal horoscopes!).

In any event, in quixotic search for a mythic city, Philby set forth with his caravan of camels and his retinue of Arab guides, who no doubt muttered much about the oddity of their quest, the quantity of their pay, and the quality of the food. Disappointingly, Philby found only some circular craters, which he believed to be of volcanic origin, with rims full of great chunks of glass. Small black spherules of glass were common. But by an amusing coincidence, Philby carried with him on his expedition a copy of the *Geographical Journal* which happened to have in it Maclaren's article on the Lake Bosumtwi crater. Reading that article changed Philby's opinion of the origin of the Wabar crater from the volcanic to the meteoritic interpretation. Glass collected by Philby around the craters was later found to contain heavily shocked, high-pressure forms of quartz and tiny droplets of nickel-bearing iron. And finally, Philby retrieved a probable iron meteorite, more the size of a camera than a camel, which was in due course analyzed and found to be meteoritic metal.

In the ensuing years the criteria for identification of impact craters have been substantially refined. The tilting of surface layers upward and outward under the crater rim, the presence of shocked mineral grains and minerals that require extremely high pressures for formation, the presence of meteoritic debris, the presence of roughly conical structures (called shatter cones) in the local rock that are produced by very rapid failure under extremely high shock loads, and the presence of impactite glass are the most diagnostic features of an impact origin. The present-day search for impact features is assisted by a wide range of laboratory techniques that did not become available until very recently. Not surprisingly, the number of authenticated impact features on Earth has risen rapidly. Over two hundred are now known.

Now that there is ample evidence for numerous impact craters on Earth, it is necessary to ask what effects these impacts have and how frequently they occur on Earth. To do so requires both understanding the effects of very large explosions and collecting all the information we can on the effects of impacts on other planets.

3

STEALTH WEAPONS FROM SPACE

Early in the ninth hour of the morning of June 30, a very unusual natural phenomenon was observed here. In the village of Nizhne-Karelinsk (200 versts north of Kirensk) in the north-west sky very high above the horizon, the peasants saw a body shining very brightly, indeed, too bright for the naked eye, with a blue-white light. It moved vertically downward for about ten minutes. The body was in the shape of a cylindrical pipe. The sky was cloudless, except that low on the horizon, in the same direction as the luminous body, a small black cloud could be seen. The weather was hot and dry, and, as the luminous body approached the ground (which was here covered by forest), it seemed to be crushed to dust, and in its place a vast cloud of black smoke formed, and a loud explosion, not like thunder, but as if from an avalanche of large stones or from heavy gunfire, was heard. All the buildings shook, and at the same time a forked tongue of flame burst upward through the cloud.

All the inhabitants of the village ran into the street in terror. Old women wept; everyone thought that the end of the world was upon them.

Report in the newspaper *Sibir*, Irkutsk, Siberia, July 2, 1908

Thus began the saga of one of the most astonishing astronomical discoveries of the twentieth century: that tremendous aerial explosions can batter huge areas of the surface of Earth without forming a crater or leaving any other lasting geological evidence. Intensive study of the behavior of meteors has given us the tools to understand such bizarre events.

Most casual observers of the night sky have seen a few faint traces of light, about the apparent brightness of a star, dashing across the sky. Within the span of a second or so, a tiny particle of interplanetary dust, trapped for endless years in an orbit around the Sun, suddenly encounters Earth's upper atmosphere at a speed of 12 to 72 kilometers per second (26,000 to 156,000 miles per hour!). From Earth's perspective, each of these speedy little cosmic grains of sand carries dozens of times as much energy as an equal mass of chemical high explosives. From the perspective of the grain, the atoms and molecules of Earth's upper atmosphere strike like an incredibly hot jet of gas, hitting with the same speed as atoms in a gas with a temperature of several hundred thousand degrees. As the blast of air strips atoms from the surface of the grain, it erupts into furiously hot vapor, far hotter than the surface of the Sun. The minerals in the grain dissociate into atoms, and the atoms are stripped of many of their electrons to make an ionized and glowing gas, called a plasma, that is left behind by the plummeting meteor. This erosion of a projectile by melting and vaporization is known in the world of military missile technology as ablation. Larger meteors, of gram to kilogram size, often leave persistent trails of glowing gas, smoke, and dust, which in some cases last for many minutes. The "smoke" is a very fine silicate dust containing tiny particles condensed from the meteor vapor as it cools, mixed with larger particles that are drops of liquid rock scoured from the surface of the meteor by ablation during its fiery fall.

The density of the atmosphere traversed by the meteor increases very rapidly as it nears the ground. In fact, the density roughly doubles for each five kilometers of altitude the grain loses. A grain entering the atmosphere vertically at fifty kilometers per second sees the intensity of the air blast double ten times (increase by a factor of over one thousand) in a single second. The pressure of the blast of hot air not only ablates the meteor, but also decelerates it. The decelerating force, caused by the pressure of the air cap in front of the meteor, can be larger than the crushing strength of the meteor material. That force increases with the square of the velocity and is proportional to the density of the air through which the meteor is passing. The density of air near Earth's surface is so great that strong stony bodies traveling at speeds of, say, twenty kilometers (twelve miles) per second are crushed to a powder by the aerodynamic pressure. Our normal intuition would suggest that the faster a projectile is moving, the more deeply it will penetrate its target. But the exact opposite is true of meteors: the fastest-moving meteors produce the highest pressures, and are crushed to dust at higher altitude. The word *meteor* is in fact reserved for objects that do *not* survive passage through the atmosphere. A solid body that reaches the ground intact is called a *meteorite*. Thus despite its name Meteor Crater was certainly made by a meteorite, not a meteor.

Photographic tracking of the flight of meteors shows that they are usually visible only while they are at high altitudes. Most of the brighter meteors burn more brilliantly as they penetrate the atmosphere, then erupt suddenly into a "terminal flare" of greatly enhanced luminosity before abruptly disappearing. It is also not rare to see "sparks" given off by a meteor during atmospheric entry. Both the sparks and the terminal flare are due to crushing (fragmentation) of the meteor by aerodynamic forces. Since we know the atmospheric density at different heights, and since we can measure the speed of the entering meteor very precisely by means of photographic tracking, we can calculate the conditions experienced by the meteor at the altitude at which it disintegrates and makes the terminal flare. We find that most meteors disintegrate under very low aerodynamic pressures, only 0.01 to 1 atmosphere, often at altitudes of 60 km (36 miles) or higher. Most meteors are weak, easily crushed material.

The hot gases ablated from a meteor glow brilliantly. As much as several percent of the total energy of an entering meteor is radiated as light and heat. Since the luminous trail of the meteor is a mixture of meteor vapor and well-known atmospheric gases, it is often possible to deduce the composition of the meteor from the spectrum of light emitted by its trail. The results are not terribly surprising in one sense: the meteor body is dominated by normal, abundant rock-forming elements, the same that form the terrestrial planets and meteorites. The very weakest meteors also contain large amounts of carbon, far more than we see in even the most carbon-rich meteorites.

It is startling but true that no meteorite fall has ever been associated with a meteor shower. Even when meteorites chance to fall during the time of a meteor shower, their orbits are unrelated to the orbit followed by the meteors. Therefore meteors are not just little particles of meteorite materials: both their physical weakness and their carbon content reveal that they are something different. But then, if meteors and meteorites are different, where do they come from? Is there some relationship between them? Are there any very large meteors?

Meteors, from the size of a pinhead (a faint naked-eye meteor) to a metric ton or so (a brilliant fireball as bright as the full Moon), fragment and vaporize high in the atmosphere. Since most meteors follow orbits associated with the orbits of known (or recently deceased) comets, it is reasonable to conclude that cometary debris is intrinsically weak. Spectroscopic studies of comets show that they are dominated by "dirty snow." Most of their mass is ordinary water ice, and the rest is carbon-bearing rocky dust. Studies of the motions of meteors show that cometary debris follows orbits that encounter Earth at very high speeds, often from 30 to 55 kilometers per second, and sometimes as high as 72 km/s. Thus their impact with the atmosphere is exceptionally violent compared with that for asteroidal debris, and they are subjected to extremely severe aerodynamic crushing forces. The origin of the large, dense, solid bodies that survive as recoverable meteorites must be different: their velocities at encounter with Earth are often below 20 km/s, and sometimes as low as about 11 km/s. They are fragments of "parent bodies" that not only have different chemical and physical properties than comets, but also pursue different orbits around the Sun.

If you've ever spent a sultry August evening watching the sky, you've probably noticed that meteors like to travel in company. On nights when meteors are rare, they seem to follow random and unrelated paths across the heavens. But on those nights when meteors are common, most appear to radiate from a particular point in the sky, fixed with respect to the stars. This point, called the radiant, is merely an effect of perspective: the associated meteors are moving together in a cluster, with nearly identical velocities and orbital paths, and Earth just happens to be in the way.

Because any fixed point in the sky lies within the boundaries of a constellation, it is convenient to name meteor showers after the constellation from which they appear to emanate: Geminids, Perseids, Leonids, Aquarids. When more than one radiant occurs in a given constellation, they are distinguished by the naked-eye star that is nearest to the radiant (Eta Aquarids, Iota Aquarids, etc.) or by the month of their annual appearance (January Bootids, June Bootids, etc.). The random background meteors that do not belong to discrete meteor showers are called sporadic meteors.

The timing of showers is highly predictable: from year to year meteor showers occur on the same dates. Some meteor showers last just one or two nights, to disappear completely until the same two nights the next year. Others, especially the weakest showers (those with the fewest visible meteors per hour), may persist two weeks or more. These meteor streams are older, and have had more opportunity to drift apart under the effects of disturbing forces, such as deflections by the gravitational attraction of the planets whose orbits they cross. Some of the older meteor streams are split into two or more branches with closely similar but distinguishable orbits. Other streams follow very closely the same orbit, with a consistent radiant from year to year, but the number of meteors seen may vary dramatically from year to year. This suggests to us that the distribution of meteors around the orbit is very clumpy, probably because these meteor streams are young and not yet widely dispersed by perturbing influences.

Some of the more prominent meteor showers have been known for centuries. Thorough searches of old records from all over the world have revealed the continuity of some streams for over a thousand years. One of the perennial streams that sometime rises to astonishing heights of activity is the Leonid shower. Reports of this shower in the older records almost invariably refer to those rare occasions when the rate of appearance of visible meteors rises to a hundred or more times the rate seen in "off years." The earliest of the known Leonid meteor storms dates from A.D. 585. Other prominent displays were recorded in 902, 934, 1002, 1202, 1366, 1582, 1602, 1698, 1799, 1833, 1867, 1901, 1932, and 1966. The yearly pattern of brilliant displays clearly reveals a repeating interval of about 33.2 years. The actual date of the peak display has shifted slowly from year to year, at an average rate of about fifteen days of offset from A.D. 902 to 1602, or about a half an hour per year. This drift, due in part to orbital changes of the meteor stream and in part to shortcomings of the calendar itself, is so slow that it never confuses the recognition of a stream. After the uniform adoption of the Gregorian calendar throughout western Europe the calendar was offset to about ten days later, and with a different average rate of drift caused by the different ways in which the two calendars accommodated leap years.

The next Leonid peak is due on November 17, 1999. The tendency of the Leonids to (approximately) commemorate centennial and millennial years will undoubtedly be much discussed as the end of the current millennium approaches. It is impossible to guess whether the pattern of recurrence we now see dates back before the birth of Christ; nonetheless, the clumpiness of the Leonid stream clearly shows its comparative youth.

The orbit of the Leonid shower has been determined from both photographic and radar observations. Their mean distance from the Sun is close to ten times as far as Earth's mean distance from the Sun (which is defined as one astronomical unit, or 1 AU). The Leonids pursue an extremely elongated elliptical orbit that is so greatly inclined relative to the plane in which the planets move that the Leonids approach Earth almost head-on in its orbital motion around the Sun. Their point of closest approach to the Sun (perihelion) is at a distance of 0.983 AU from the Sun, just inside Earth's orbit. If the perihelion distance of the swarm were an equal distance *outside* Earth's orbit it would never intersect Earth and we would be unaware of its existence.

The comet Tempel-Tuttle follows an orbit that is strikingly similar to that of the Leonids. Its perihelion distance of 0.982 AU is virtually identical to that of the Leonid swarm. It appears that the Leonids contain a dense cloud of dust released during recent stressful passages of comet Tempel-Tuttle by the Sun. But comets are mostly ice: it is the evaporation of ices, under the intense heating of the Sun, that causes the glowing clouds of gas and streams of dust to spray outward from the frozen comet nucleus. Larger pieces of dirty cometary ice may also be dislodged by this furious evaporation.

The orbits of many other meteor swarms have been determined. Of the eighty known reasonably reliable showers (some are quite weak) at least twenty, including the brightest ones, have orbital connections with known comets. Many meteor streams follow orbits that are very elongated and very highly inclined, making it impossible to associate them with any other solar system bodies except long-period comets. Almost all long-period comets have orbital periods of one hundred thousand years or more. They therefore cannot have completed an entire orbit within historical time, and almost certainly have not passed through the inner solar system since the dawn of recorded history. If there is any trace of their existence in human records, it must be in the most shadowy recesses of ancient mythology.

Interestingly, three regular meteor showers have orbits connected with three asteroids whose orbits bring them very close to Earth; Icarus, Phaeton, and Hermes. All three of these near-Earth asteroids have highly elongated (eccentric) orbits, more typical of short-period comets than other asteroids. Icarus is associated with the Arietid meteor shower, Phaeton is the apparent source of the Geminid shower, and Hermes is connected with the October Cetids. It it likely that these asteroids are simply the bare cores of extinct short-period comets, that have been near the Sun so long that ices have been nearly completely lost from their outer layers. Several physicists, beginning with astrophysicist-turned-science-fiction-writer David Brin, have developed computer models of the evaporation of icy materials from comet nuclei as they pass near the Sun. The general conclusion of these studies appears reliable: a comet in such an orbit would lose all the ices from the outermost meter or so of its surface. The ice-rich interior would be so well insulated by a layer of dust left behind on the surface by the evaporating ices that the ices could last practically forever; for 100 million to a billion years or more.

A hundred million years is a very long time for a near-Earth asteroid (NEA). These bodies pass through the heavily trafficked inner solar system; many cross the orbits of three or four different planets. The average NEA crosses the plane of the planetary system about once per year. Each NEA passes within 100,000 km of a terrestrial planet about once per million years, on average, which is close enough to stir up its orbit around the Sun. A passage 100,000 km from Earth would be only a quarter of the way from Earth to the Moon. That's close! Because of the recurrent hazard of collision with Earth, Venus, Mars, Mercury, or the Moon, NEAs rarely survive much more than 100 million years before catastrophe overwhelms them. Those that are not destroyed by planetary impact may wander into Jupiter-crossing orbits and be hurled by Jupiter's immense gravity completely out of the solar system, or even be thrown into the Sun. NEAs are accidents waiting to happen.

The near-Earth asteroids and comets have very short lifetimes. Because we see thousands of them, there must be a continuing source of new ones. Studies of the orbital evolution of comets and asteroids by a number of astrophysicists have shown that there are two plausible sources of new NEAs. First, asteroidal bodies from the heart of the asteroid belt may wander into orbits whose orbital period is harmonically related to Jupiter's orbital period, a situation colorfully described as an orbital resonance. The gravitational effects of Jupiter then recur repeatedly and build up to so large an effect that resonant asteroids can be kicked into eccentric orbits that carry them repeatedly through the inner solar system. Such disturbed asteroids typically have orbits that are only moderately eccentric and inclined. Second, fresh long-period comets that pass close to Jupiter may be perturbed by Jupiter's gravity into short-period orbits with low inclinations but much higher eccentricities. Neither source is by itself sufficient to provide the needed supply rate of NEAs. The computed rates of injection of new NEAs for the two mechanisms are roughly equal. Interestingly, the *total* of the two is just what we need! Thus we are led to the conclusion that about half of the NEAs are extinct comet nuclei and half are errant belt asteroids.

But for our present purposes, this means that about half of the NEAs have the chemical and physical properties of ice-rich comets, and about half are fragments of rocky or metallic bodies from the asteroid belt. These different materials surely have very different strengths and behave very differently during entry into Earth's atmosphere. In addition, pieces of cometary surface material from which the ices have been lost by evaporation may be very fluffy, porous aggregates with very little strength. Indeed, the 1986 spacecraft missions to Halley's comet found that the entire nucleus has a density only 20 percent of that of water: it must be 80 percent empty space, a vast network of cracks and voids.

If most of the particles in meteor streams arise from comets, then we can easily enough picture how these streams came into existence by observing fresh, active comets as they pass by. Such comets, on their first few passes through the inner solar system, "light up" as they approach the Sun. Ices evaporated off the surface of the solid comet nucleus by solar heating expand outward from the dayside of the nucleus, blowing along all the smaller dust particles. A vast cloud of gas and dust expands in a roughly spherical cloud around the comet until it gets far enough away so that other forces dominate the motion of the gas and dust. The gas becomes so rarified that the dust ceases to heed its presence: gas-dust collisions become so rare that the dust orbits the Sun nearly at the same speed, and nearly in the same orbit, as the comet nucleus. The dust emitted from the nucleus at first forms a planet-sized dense cloud around the nucleus, but the small velocity difference produced by the expansion of the gas-dust cloud causes the dust grains to follow orbits with slightly different velocities, and hence with slightly different periods. The dust cloud begins to smear out along the orbit of the comet. Over many years, and many searing perihelion passages, the dust band eventually extends all the way around the comet's orbit. The effects of radiation pressure from the Sun help smear out the dust cloud, making it fill an ever thicker doughnut-shaped (toroidal) volume about the orbit of the comet.

The nucleus of the comet also changes in response to this process: large dust particles, too big to be blown away by the evaporating ices, accumulate on the surface of the nucleus to make what is called a lag deposit. When this layer of fluffy, black dust gets thick enough, according to the work of Brin and others, it becomes an excellent shield against solar heating. Sunlight cannot penetrate the layer, and heat cannot be conducted through it. The ices a meter or so beneath the surface lose all contact with the Sun, and no longer evaporate madly around the time of perihelion passage. Of course an impact of a small meteorite can open a hole in this dust layer and permit "fountaining" of gases where the sunlight now can penetrate to the ice. Indeed, photographs of the nucleus of Halley's comet taken by flyby spacecraft from the European Space Agency and the USSR in 1986 show that only a few percent of the surface of the nucleus is in eruption near the time of maximum solar heating. The rest is covered by an extremely black dust deposit.

As long as a comet nucleus is still active, fresh clouds of dust are emitted at each perihelion passage. After a number of centuries, a broad, diffuse cloud of "old" dust will stretch all the way around the comet's elliptical orbit. Earth may take weeks to pass though this extended cloud, while experiencing a prolonged but very weak meteor shower with either a poorly defined radiant or multiple, rather closely spaced radiants. Embedded in this diffuse meteor stream may be one or more very dense, small clouds of "young" dust that extend only a few percent of the way around the orbit. The phasing of the motions of Earth and a dense dust clump will in general permit very intense meteor displays at infrequent but rather regular intervals. The youngest, densest clumps of dust will have orbits that most closely resemble the current orbit of the comet nucleus from which they arose.

Encounters of Earth with dense dust swarms produce spectacular meteor storms. One such storm, a shower in November 1799, was seen only in a limited area in Central America and Venezuela. Since the shower of this date pursues an orbit about the Sun that causes it to appear (from Earth!) to emanate from the direction of the constellation Leo, it is referred to as the Leonid meteor shower. News of this event did not reach Europe in time to contribute to (or complicate!) the contemporary debate about the extraterrestrial nature of meteorites.

However, shortly after the general acceptance of the reality of the fall of stones from the sky, in 1833, the heavens revealed a most extraordinary phenomenon. On the night of November 13, the date of the well-known but modest annual Leonid shower of meteors, a spectacular meteor storm occurred. Instead of a modest streak of light every few minutes, it rained meteors, as if the stars were falling from the sky, too numerous to count. The Leonids usually are typical naked-eye meteors, about the size of a grain of sand, which become incandescent and vaporize when they collide with Earth's atmosphere at immense speeds. But in 1833 there were many brilliant fireballs that left trails of dust and smoke. In South Carolina, the *Charleston Courier* reported "a meteor of extraordinary size was observed at sea to course the heavens for a great length of time, and then explode with the noise of a cannon." Some observers reported sulfurous odors; others waxed fortean with reports of a substance "like lumps of jelly," "the white of an egg made hot," "soft soap," "boiled starch," et cetera. Numerous reports have linked the discovery of white or gray viscous, volatile, combustible material with brilliant fireballs, usually in meteor storms. There are even rare reports of large lumps of dirty ice falling from clear skies.

A tutor at Yale who witnessed the 1833 Leonid storm was impressed both by the implicit threat and by Earth's astonishing defense against it. Alexander C. Twining made his observations in 1834 in the *American Journal of Science*:

> The multitude of bodies was such as no man can venture with confidence to limit by numbers: and, had they held on their course unabated for three seconds longer, half a continent must, to all appearances, have been involved in unheard-of calamity. But that Almighty Being who made the world, and knew its dangers, gave it also its armature—endowing the atmospheric medium around it with protecting, no less than with life-sustaining properties: and, considered as one of the rare and wonderful displays of the Creator's preserving care, as well as the terrible magnitude and powers of his agencies, it is not meet that such occurrences as those of Nov. 13th, should leave no more solid and permanent effect upon the mind, than the impression of a splendid scene.

(Such sentiments are rarely encountered in the scientific journals of the late twentieth century. It is by no means obvious to me that this development is an improvement.)

In 1966 the Leonids returned. The display occurred at night in western North America, but almost perfect cloud cover on the night of November 16 hindered observations. I recall passing several boring hours in the California desert under dense, unbroken clouds before giving up. But a team of thirteen amateur observers atop Kitt Peak in Arizona were lucky enough to have the clouds part to reveal to them the display of a lifetime. Dennis Milon, one of the Tucson observers, related to me that, as they drove up Kitt Peak around midnight, the skies began to clear. They counted a respectable 33 Leonids in the hour between 1:30 and 2:30 A.M. in a clear sky. Between 2:50 and 3:50 one observer recorded 192

Leonids, of which 30 were as bright as the brightest stars (negative magnitudes*).

By 4:10 A.M. the count was up to 30 *per minute.* A fireball of approximate magnitude −8 (similar to a first-quarter moon) then exploded overhead, and the skies almost literally began to rain light. Milon reported that "by 4:30 there were several hundred per minute. At 4:45 the meteors were so intense we guessed how many were seen by a sweep of the head in one second. The fantastic rate of about 40 per second was reached at 4:54 A.M. . . . Some of the brighter ones left trains for several minutes, and were photographed. Fully half left trains. We took many photos. . . . a forty-three-second exposure of the Big Dipper . . . shows 43 Leonids. . . . The view was so spectacular that we didn't know where to look. . . . looking directly at the radiant gave the effect in depth of the Earth moving through space. . . . Sometimes we would spin around, taking in the whole sky. Or we alternated with looking toward the western horizon (where it was very clear) and gazing right at the radiant. Different parts of the sky would light up, and we would glance here and there. Everyone was yelling and laughing at the incredible, dazzling sight, and at our luck in seeing it."

The meteor count rate was over 500 per minute for almost exactly one hour, peaking at about 2,500 per minute (150,000/hr). These few lucky observers had witnessed the most spectacular meteor shower in recorded history, surpassing even the 1833 Leonids. And that meteor storm is in an orbit that brings it back every thirty-three years.

It is not too late for you to make your own plans for the night of November 16–17, 1999. . . .

Other spectacular meteor storms have occurred in recorded history, and it is no surprise that they often gave occasion for rumors of the end of the world. A brilliant shower was seen on the night of November 27, 1872, when Earth passes very close to the orbit of Biela's comet, which had split into two major (and, evidently, countless minor) pieces on January 13, 1846, and altogether vanished by 1865. This Andromedid shower was seen again on the same date in 1885, but it is now reduced to a feeble annual display as the bodies

---

*The apparent brightness of stars and other heavenly bodies is described on an ancient scale in which the brightest stars, such as Vega, are about "first magnitude" and the faintest stars visible to the unaided eye are "sixth magnitude." The lower the number, the brighter the object. Bodies brighter than Vega must then be assigned negative magnitudes: the full Moon is about −12.5 magnitudes and the Sun is a scorching −26.7 magnitudes. Fireballs with apparent brightnesses of −15 or so, brighter than a full Moon, are reported every year.

. . . .

in the shower have dispersed themselves all the way around the comet's orbit. Other brilliant meteor displays with the same date and radiant, evidently also associated with an earlier breakup of Biela's comet, were seen in the years 524, 585, 837, and 899. The Leonids are a young, clumpy swarm that is strongly concentrated in space. Although some Leonids are seen each year in mid-November, spectacular meteor storms seem to recur at thirty-three- or thirty-four-year intervals, when Earth intersects their orbit at the time when the clump of meteoric material passes by.

The Leonids and Andromedids are not especially noted for their displays of very brilliant fireballs, but some other streams are. Historical surveys of meteor data reveal that another prominent November meteor shower, the Taurids, was once extremely conspicuous for its numerous brilliant fireballs. During the eleventh century the brilliant Taurid fireballs were the most notable meteoric phenomenon of all. At that time the two modern active branches of the stream, with radiants about 16° apart, were both prominent, suggesting derivation from two distinct cometary bodies in similar orbits. From about 1100 to 1869 the Taurids were lost, or so inconspicuous as to evade detection. November of 1920 brought another brief flurry of bright fireball activity. In the 1960s careful orbital studies of the Taurids revealed that there are indeed two distinct streams, but with essentially identical orbital planes. Thus the case for a common origin was greatly strengthened. The orbits suggest, but do not prove, that at least two major ejection events from Enke's comet, in the fifth century A.D. and around 2700 B.C., were responsible for the Taurids. Enke is unusual in that it is the active comet with the shortest known orbital period. Because of the predictability of these annual showers, it is conceivable that Anaxagoras's real feat was the prediction of the date of a display of brilliant fireballs that just happened to drop a meteorite, or that accidentally coincided with the date of fall of an asteroidal meteorite. No cometary meteor shower has ever produced a meteorite fall.

Enke is also the parent of the particularly elusive Beta Taurid meteor shower that spans the entire time from about June 5 to July 18, with a peak near July 1. Since this shower occurs on the dayside of Earth it is generally observed only by radio techniques (usually, by bouncing radar pulses off the ion trails left by the meteors as they burn up). This very broad and old stream crosses the orbits of all the terrestrial planets nearly in the plane of the ecliptic, thus providing endless opportunities for dynamical complexity and collisional mischief. Encounters of this stream with Earth therefore can occur at two distinct, widely separated times of year. The combination of all of Enke's family members, including the north and south Taurid streams of November and the Beta Taurids of June, is appropriately known as the Taurid complex. The cometary events responsible for these multiple episodes of fragmentation are clearly important to the terrestrial planets. We shall return to this problem in chapter 11.

Halley's comet, the most active comet of the twentieth century, is also associated with two showers at different times of year, the Eta Aquarids of May and the Orionids of October.

Yet another erratic stream is the Lyrids, which can be seen any year as a faint display lasting from April 16 to 25. But on April 19, 1803 a spectacular display was observed in the southeastern United States. Careful searches through older records revealed brief, intense earlier displays associated with the same stream on April 9, 1095, April 10, 1096, and April 10, 1122. By 1861 the orbit of the Lyrids had been linked to that of the comet Thatcher. This linkage permitted calculation of the cometary orbit back to a brilliant meteor display on March 16, 687 B.C. The evidence suggests a very lumpy distribution of material about Thatcher's orbit, again with a small number of recent events superimposed on a background of material from much older events.

Frequent breakup and erosion of cometary nuclei provides not only vast quantities of meteoric dust, but also occasional larger fragments. It is interesting to reflect upon the possibility that a meteorite-sized chunk of cometary material might survive its fall to Earth. Perhaps the most intriguing report of such a possible fall was reported in April 1995. A Chinese geologist, Zhong Gongpei, responded to a report by several farmers that they had just seen several large chunks of ice drop from cloudy streaks in the sky into a rice paddy. Zhong quickly retrieved a fist-sized chunk and rushed it to a frozen-food warehouse for safekeeping. Chinese meteorite experts are reportedly already examining the object.

One of the stranger pieces of lore regarding large meteors come from an unlikely source—the seismometers left on the surface of the Moon by the Apollo astronauts. During late June of 1975, over a five-day period, as many ton-sized impactors struck the Moon as in the entire preceding five years. The dates correspond to the heart of the daytime Beta Taurid shower. If a number of similar bodies had hit Earth during the same few days, they would have entered in daylight, and would have had to compete with the Sun to be seen.

. . . .

Another unique lunar event was recorded on A.D. June 25, 1178 by Gervase, a monk at Canterbury, England:

> *In this year, on the Sunday before the feast of St. John the Baptist, after sunset when the Moon had first become visible, a marvelous phenomenon was witnessed by some five or more men who were sitting facing the Moon. Now there was a bright new Moon, and as usual in that phase its horns were tilted toward the east; and suddenly the upper horn split in two. From the midpoint of the division a flaming torch sprang up, spewing out, over a considerable distance, fire, hot coals, and sparks. Meanwhile the body of the Moon which was below writhed, as it were, in anxiety, and, to put it in the words of those who reported it to me and saw it with their own eyes, the Moon throbbed like a wounded snake. Afterwards it resumed its proper state. This phenomenon was repeated a dozen times or more, the flame assuming various twisting shapes at random and then returning to normal. Then, after these transformations, the Moon from horn to horn, along its whole length, took on a blackish appearance. The present writer was given this report by men who saw it with their own eyes, and are prepared to stake their honor on an oath that they have made no additions or falsifications in the above narrative.*

Meteorite expert Jack Hartung has interpreted Gervase's narrative as referring to the formation of the crater Giordano Bruno on the Moon by a comet or asteroid impact. Photographic mapping has revealed that Bruno is probably the youngest large crater on the Moon. Assuming that Hartung's identification is correct, then Bruno's diameter of about 20 km (12 miles) implies an explosive impact on the Moon in A.D. 1178 with an energy of about 120,000 megatons of high explosives. Imagine the blast damage from a bomb containing 24 tons of high explosive, such as TNT or RDX, larger than the biggest chemical bomb ever used, the famous Blockbuster of World War II. Now imagine 5 billion people, the entire population of Earth, each setting off a 24-ton explosion at the same time. That is 120,000 megatons. All this energy would be carried by two cubic kilometers of rock traveling at 14 kilometers per second. For comparison, an all-out nuclear exchange involving all the world's nuclear weapons would deliver no more than 20,000 megatons.

Several features of Gervase's document deserve mention. First, the date in late June (June 18 in the Julian calendar; June 25 in the modern calendar) again strongly implicates the Taurid complex. Second, the year is also telling: this was the century in which the Taurid fireball flux was at a maximum. Finally, there is the interesting fact that the impact occurred on the Moon, not on Earth. Consider a 2 km³ rock approaching the Earth-Moon system at about 14 kilometers per second. It is randomly targeted at the Earth-Moon system. The Moon presents a circular target with a radius of 1,740 km, and Earth presents a circular target with a radius of 6,370 km. But there is more to the picture than simple geometry! Both Earth and Moon have gravitational fields that allow bodies that would have missed them without their gravitational attraction to hit them. Slow-moving projectiles are much more strongly diverted because they spend more time near their putative targets, and of course more massive targets are better at deflecting projectiles of any speed. Assuming a 14-kilometer-per-second projectile, the focusing effects of Earth's and the Moon's gravity make it more than twenty times more likely that a cosmic bullet of Taurid speed should strike Earth than that it should hit the Moon. Earth was extremely fortunate to receive no more than a rain of fireballs during the twelfth century!

The saga of the Taurids is, however, not complete without returning to the event described at the opening of this chapter. The June 30 fireball that startled and terrorized residents of many villages in Siberia in 1908 was seen to descend toward the northern horizon, into an area with a very sparse population of nomads, most of whom spoke only their native language, not Russian.

A close eyewitness of the fireball, an Evenok herdsman named Ilya Potapovich Petrov, was near the site of the explosion in the valley of the Tunguska River. His brother, who spoke no Russian, was perhaps the closest witness. Ilya's recounting of his brother's story, as relayed by a Russian interviewer, is the closest we can get to any eyewitness account:

> *One day a terrible explosion occurred, the force of which was so great that the forest was flattened for many versts along the banks of the river Chambey. His brother's hut was flattened to the ground, its roof carried away by the wind, and most of his reindeer fled in fright. The noise deafened his brother, and the shock caused him to suffer a long illness. In the flattened forest at one point a pit was formed from which a stream flowed into the river Chambey. The Tunguska road had previously crossed at this place, but it was now abandoned because it was blocked, impassable, and*

moreover the place aroused terror among the Tungusk people. From the Podkamennaya Tunguska River to this place and back was a three-day journey by reindeer. As Ilya Potapovich told this story, he kept turning to his brother, who had endured all this. His brother grew animated, related something energetically in the Tungusk language . . . striking the poles of his tent and the roof, and gesticulating in an attempt to show us how his tent had been carried away.

The night after the explosion the skies over eastern Europe glowed brilliantly. There is a tale that the garrison of a Polish army base deserted en masse, believing it was the end of the world, and joined a monastery. Even as far away as England the glowing dust and gas from the explosion was evident. One witness wrote to the London *Times* on July 1, 1908 of her experience of the previous night:

Sir,—I should be interested in hearing whether others of your readers observed the strange light in the sky which was seen here last night by my sister and myself. I do not know when it first appeared; we saw it between 12 o'clock (midnight) and 12:15 A.M. It was in the northeast and of a bright flame-colour like the light of sunrise or sunset. The sky for some distance above the light, which appeared to be on the horizon, was blue as in the daytime, with bands of light cloud of a pinkish colour floating across it at intervals. Only the brightest stars could be seen in any part of the sky, though it was an almost cloudless night. It was possible to read large print indoors, and the hands of the clock in my room were quite distinct. An hour later, at about 1:30 A.M., the room was quite light, as if it had been day; the light in the sky was then more dispersed and was a fainter yellow. The whole effect was that of a night in Norway at about this time of year. I am in the habit of watching the sky, and have noticed the amount of light indoors at different hours of the night several times in the last fortnight. I have never at any time seen anything like this in England, and it would be interesting if anyone would explain the cause of so unusual a sight.

Yours faithfully,
Katharine Stephen
Godmanchester, Huntingdon
July 1, 1908

The year of the Siberian explosion was a time of civil turbulence in Russia. In 1905 more than a thousand peaceful demonstrators had been killed when the police opened fire on them before the Winter Palace in St. Petersburg. That fall a general strike was called, largely instigated by socialist and communist agitators, which resulted in a decree by Tsar Nicholas II establishing a popularly elected parliamentary body, the Duma. The Duma, packed by radicals, was dissolved by the Tsar in 1906. A second election was held and a new Duma convened in 1907, but it was even more polarized than its predecessor. This parliament was also dissolved by decree, and a third Duma was elected and convened in the fall of 1907. The elections were managed so as to assure domination by conservative elements who wished to preserve the status quo, and reformist elements were left frustrated and with no real power. Prime Minister Stolypin was murdered by radicals in 1911. Meanwhile, the Triple Entente of England, France, and Russia, formed in 1907, set Russia on a path to involvement in World War I in 1914, perhaps the most important proximate cause of the Revolution of 1917, the fall of the Romanovs, and the establishment of a Communist state. In the midst of this turmoil, it is scarcely surprising that no quick investigation of the Siberian fireball was attempted.

Geologists working in the area in 1924 interviewed many of the surviving witnesses of the events of sixteen years earlier, and interest in the event spread. The sounds of the fireball had been loud over an area of at least a million square kilometers; hundreds of witnesses had seen the fireball in the sky and marked its direction. Hundreds reported that it fell to earth north of them, and a tremendous explosion ensued. And of course millions in Europe noticed that the skies were uncannily bright. Seismic records were assembled showing a powerful explosion in the air (the first networks of seismic stations had been installed only about ten years earlier), and microbarograph records of the atmospheric pressure waves from the blast were traced for three complete circuits of the world (the first microbarographs had been installed only five years earlier). In full expectation of finding an impact crater and a large mass of meteorite material, Leonid A. Kulik set out from Leningrad in March of 1927 to explore the Tunguska basin.

As Kulik's expedition approached the target area, they began to encounter clumps of seared trees on elevations. The last blizzards of winter, alternating with days of thaw that turned the ground into a quagmire, slowed the expedition to a few kilometers per day. Finally

Kulik reached the center of the site on snowshoes. What he found was stunning.

Over an area of at least 2,000 square kilometers, roughly 30 km (18 miles) in average radius, the forest had been felled radially outward from a central point. Close to the center the trunks of the trees remained standing upright, but all their branches had been stripped off and thrown to the ground. Every tree was scorched or charred by a brief exposure to extremely intense heat. But, incredibly, there was no crater to be found, no meteorites to be collected, no impactite glasses. The damage was breathtaking in its extent, but the evidence of the perpetrator was shockingly fragile. A century later the forest would be regrown, the downed trees rotted away to nothing. A visitor would never know that this had been the scene of a stupendous explosion. After a century, the only surviving evidence would be vast quantities of tiny spherules of magnetite, the condensed residue of oxidized meteorite material from the fireball. But who would think to look for them there?

The place of fall of the Tunguska object was of course random. By good fortune, it struck in a very remote area, where there were no more than twenty people within fifty kilometers of ground zero. All of those were injured, and two were reportedly killed. Had it struck at the same latitude four hours and fifty-two minutes later it would have hit St. Petersburg and its hundreds of thousands of residents. An offset of a few minutes more would have threatened Helsinki. Yet later, and it would have hit Stockholm. A delay of six hours and eight minutes would have targeted Oslo instead of desolate Taiga.

Most random events, of course, would be over the oceans, and would almost certainly not be seen. What kind of body created the blast over the Tunguska valley? How can we account for the remarkable phenomena of this event? And finally, how often do Tunguska-like explosions occur? The detective work needed to answer these questions would take us deep into the twentieth century, into the era of nuclear weapons. The effects of such large explosions are explored in chapter 4.

4

BRIGHTER THAN A THOUSAND SUNS

In the ship channel of Halifax harbor, just off Dartmouth, Nova Scotia, the French cargo ship *Mont Blanc*, laden with 2,300 tons of picric acid, 200 tons of TNT, 35 tons of benzole, and 10 tons of guncotton bound for the Great War, caught fire and exploded with the force of nearly 3 kilotons of TNT. A powerful shock displaced the ground outward from the center of the blast, shearing off utility poles and jerking the foundations out from under buildings. A moment later the air blast struck, demolishing the waterfront structures and crushing homes and businesses more than two miles from the ship. Windows, shattered into glass spears and accelerated to half the speed of sound by the blast wave, riddled the interiors of buildings with deadly shrapnel. A tidal wave driven by the blast raced into the harbor, lifting thousand-ton ships out of the water and hurling one onto the shore. Fires by the hundreds, ignited by overturned stoves and furnaces and downed electric power lines, sprang up in the ruins. Church bells sixty miles away rang as the ground and air shook them. Heavy rocks, dredged from the bottom of the harbor and thrown into high trajectories by the blast, slammed into the ruins and the dazed survivors. Over a thousand tons of red hot metal shrapnel from the exploded ship rained down with the rocks from a black sky. Distant observers, hearing the greatest man-made explosion before the invention of the atomic bomb, watched in astonishment and alarm as a towering mushroom-shaped cloud rose three miles high over Halifax. When the casualties were tallied, 1,963 bodies were counted. Over 9,000 were injured, of whom 199 were blinded by flying glass. Hundreds more remained forever missing, all traces of them destroyed by the blast and fires.

Sailors mobilized to search for survivors wandered through the ruins in a daze. An anonymous sailor, overwhelmed by the scene of devastation and carnage, was overheard to mutter, "Looks like the end of the whole bloody world, don't it?"

HALIFAX, NOVA SCOTIA
Events of December 6, 1917

Remember this scenario: this is what happens when a three-kiloton explosion occurs in a populated area.

Not content with the level of destruction afforded by chemical explosives, and goaded in part by the specter of a German atomic bomb launched on transoceanic rockets, American, British, and Allied scientists during World War II developed the first nuclear explosives. With the fall of Germany in early 1945, the prospect of an enormously bloody inch-by-inch invasion of the Japanese home islands loomed. At the Trinity test site in New Mexico, on July 16, 1945, a nuclear device with an explosive yield of nineteen kilotons (nineteen thousand tons of TNT) was detonated atop a tower in a remote location in the Jornada del Muerto desert. J. Robert Oppenheimer, observing from a bunker, quoted the Bhagavad Gita: "brighter than a thousand Suns." Because of the intense time pressure surrounding the test, almost no data on the effects of a nuclear blast on buildings, vehicles, and living organisms were collected. The question was whether the bomb would work: it worked.

On August 6, just three weeks after the Trinity test explosion, an atomic bomb was dropped on the Japanese city of Hiroshima. The tail gunner on the B-29 *Enola Gay*, Bob Caron, summed up the scene: "It was a peep into Hell." Every citizen of the twenty-first century should be required to read the personal experiences of the residents of the bombed city. War is indeed hell, but nuclear war is a distillation of hell, a level deeper than any plumbed by Dante. The twenty kilotons of explosive power poured out upon Hiroshima killed at least seventy thousand people. Many more vanished without a trace. Three days later, a second nuclear weapon, of different design, was dropped on Nagasaki. The world's supply of nuclear weapons had been exhausted, bringing about the hoped-for surrender of Japan.

We cannot progress in understanding the threat of large impact events without coming face to face with the effects of large nuclear explosions. I would like to convey to you as clearly as I can the consequences of a twenty-kiloton explosion in a city, but at the same time I hesitate to subject the reader to the emotionally wrenching experience of reading the eye-witness reports of those who survived, severely burned and wracked by radiation sickness, to find everything they had known destroyed, everyone they had known dead or dying a horrible death. But it is essential for the purpose of this book that the reader understand the effects of large explosions upon actual individuals living in a real city; not damage to buildings, not cold statistics. The following brief account, edited to omit phenomena associated with nuclear radiation (which is absent in cometary and asteroidal impact events), gives the experience of a doctor, Tatsuichiro Akizuki, who lived through such a hell. It is August 9, 1945. The setting is the Urakami First Hospital on the outskirts of Nagasaki.

> *I shouted "Enemy plane! Look out! Take cover!" At the same time I . . . threw myself beside the bed. Flash! A white flash sparkled. The next moment, "Bang! Crack!" A great shock came upon our bodies, upon our heads, and upon our hospital. . . .*
>
> *Looking south-west, I was shocked. The sky is pitch dark, covered with something like cloud or smoke. Under that blackness there was, hanging over the ground, a yellow-brown smoke. The obscured ground gradually became visible, and the view rooted me to the ground in horror.*
>
> *All the buildings on Earth were on fire, both large ones and small ones with straw-thatched roofs. Far off along the valley, Urakami Church (Roman Catholic), the largest chapel in the East, is burning. The Technical School, the big two-storied wooden one, is on fire; houses and the Ordnance Factory, too. Electric poles are wrapped in flames like so many pieces of kindling. Trees on the mountains and hills are smoking. The leaves of potatoes are also smoking in the field. "Burning" is not adequate for these scenes. It seemed as if the ground was sending flames forth, wriggling and erupting. The sky is dark, the ground is scarlet, and between them, yellow smokes are hanging. Three tones of colors—black, yellow, and scarlet—are ominously overwhelming the people who are running about to escape from fire. Human beings looked like so many worms. "What is this? It is not the Urakami First Hospital that was bombed!" I could understand only that much. "This ocean of fire! This sky of smoke! I wonder if this isn't the end of the world." . . .*
>
> *At 11:00 that morning a teacher was sitting in the faculty room of Shiroyama Primary School with her back towards the windows. The gigantic pressure of the blast from the detonation struck the back windows at right angles, breaking the window-panes to pieces, and at the same time blew against her back with mighty force, driving the fragments of glass through her thin summer blouse into the muscles of her back.*

Dr. Akizuki's experiences are described in a little book entitled *Document of A-Bombed Nagasaki*. The book contains no date or place of publication, no publisher's name. It is a personal, powerful testimony by an intelligent and educated man of an experience of devastation and human suffering almost too powerful to tell. Read his testimony if you can find it, or read any of the hundreds of others in print in English. And remember: this is what happens when a twenty-kiloton aerial explosion occurs in a populated area.

In the ensuing years much larger nuclear weapons were developed. The hydrogen (fusion) bomb was first tested on November 1, 1952, detonating with the power of six megatons (6 million tons of TNT). The Soviet Union tested its own hydrogen bomb within a year, and the nuclear arms race escalated further. Intercontinental ballistic missiles (ICBMs) were built to carry these weapons over distances of five thousand miles or more. The greatest man-made explosion in history, nearly sixty megatons, was detonated on the Arctic island of Novaya Zemlya in 1962. Since then, steady improvements in the accuracy of missile guidance systems has led to a constant decrease in the size of nuclear warheads, which now typically have explosive yields of a few hundred kilotons. The nuclear testing programs of the United States, largely conducted in Nevada and on islands in the South Pacific, and of the Soviet Union, carried out in Kazakhstan and in the Arctic, have uncovered a vast amount of information on the nature and effects of very large explosions. Explosions on Earth's surface, underground, on water, deep under water, high in the air, and in space have given us a very broad understanding of the behavior of large explosions in a wide range of environments. Setting aside the phenomena related to the immediate gamma radiation from the blast, the decay of fission products, induced radioactivity in bomb debris and crater ejecta, and radioactive fallout, all of which are irrelevant to our tale, a grim roll-call of effects remains.

Impact explosions begin with the nearly instantaneous conversion of the kinetic energy of motion of the impactor into heat. Surface explosions produce a small, intensely hot fireball with enormous internal energy content and pressure. Fireball temperatures of several hundred thousand degrees are achieved briefly. At such temperatures the intensity of radiation from the fireball (the energy emitted per square centimeter per second) is millions of times as great as that emitted by the surface of the Sun. Most of the radiation from so hot a surface is not at visible wavelengths, as it is for the Sun, nor even in the ultraviolet. It is in fact near 100 Å wavelength, in what is called the soft x-ray region. Radiation of this wavelength cannot penetrate a meter of air without being absorbed, so the energy is held in near the physical surface of the fireball as the fireball expands at, initially, several times the speed of sound. The x-radiation is no threat to a bystander. If you are close enough to receive a dangerous dose, you are within the area that will be engulfed by the expanding fireball in the next millisecond: you have far worse problems than mere x-ray overdose.

The fireball drives a powerful shock wave outward in all directions. The fireball continues to expand until it reaches pressure balance with the surrounding atmosphere. By then the surface of the fireball has cooled to a few thousand degrees, but the fireball is so much hotter than the surrounding atmosphere that it is very buoyant. All the surrounding countryside, scorched by the intense heat, is now in flames. An immense bubble of hot gas begins to rise at a speed that may surpass a third of the speed of sound. Cooler air from the surrounding countryside flows rapidly inward toward ground zero to replace the rising column of hot gases, blowing at hurricane speeds and whipping the flames into a fire storm. The top of the rising column of gases further cools, expands, and mixes with surrounding air until it finally loses its buoyancy. The column of hot gases assumes the distinctive mushroom shape as it rises. The cloud from multi-megaton explosions can reach the stratosphere, which is stably stratified, where the cloud tops out but continues to spread horizontally.

During the time that the fireball is highly luminous, the total heat loads on surfaces exposed to line-of-sight view by the fireball may be very high. If radiation loads of more than about ten calories per square centimeter are accumulated by almost any common flammable material (structural wood, trees, window shades, clothing, draperies, etc.) that material will burst into flames. The experience of medical personnel in Hiroshima and Nagasaki was that severe fire burns on survivors were rare: those exposed to fire were almost without exception killed by the fire storm or by blast effects. Temperatures high enough to melt silicates and make droplets of glass are commonly found near surface bursts. Material exposed directly to the fireball will be evaporated. Small dust particles raised by the impact, and condensible rock vapors from the explosion, are entrained by the rising fireball.

Explosions set off on the surface, or very close to it, produce craters with excavated volumes that are proportional to the explosive yield of the blast. The radius of an explosion crater is proportional to the cube root of the explosive yield: a one-thousand-megaton explosion will excavate about one thousand times as much dirt as a one-megaton explosion, and produces a crater that is about ten times as large in diameter (one hundred times as large in area) and ten times as deep. A typical impact on Earth excavates a mass of material of about one hundred times the mass of the impactor. Application of these rules to Meteor Crater, thirteen hundred meters in diameter, suggests an explosive yield of about fifteen megatons. Impact craters, like surface nuclear explosion craters, tilt the strata of the country rock upward all about the rim, which is significantly raised above the previous land surface. High-speed ejected debris destroys everything within five or six crater radii of the point of impact.

The blast wave from a surface explosion affects both the atmosphere and the ground. Since the speed of sound is several kilometers per second in rock, but only three hundred meters per second in air, the ground wave travels much faster. The explosion fireball, produced by the conversion of the energy of the impactor into heat, displaces the surrounding surface material outward in all directions, driving a powerful pressure wave. As this strong shock wave races outward from the impact site, it raises a storm of dust. The first mechanical effect felt by a building a few kilometers away from the blast is a rapid displacement of the ground under it, in the direction radially outward from the explosion site. This impulse may shear the building off at ground level. Fires ignited by the luminosity of the early fireball may, if exposed at the surface, be blown out by the wind blast. In urban rubble, fires started by electrical shorts, broken natural gas lines, or exploding fuel tanks on vehicles will often be buried in loose debris, and will be unaffected by the wind blast. Widespread fires generate their own wind, which fan the flames into devastating firestorms. Further, in urban settings the blast wave will shatter glass into razor-sharp daggers and accelerate the fragments to speeds that approach the speed of sound. Building interiors on the side facing the blast will be hardest hit, but the blast wave wraps around standing structures and implodes the windows from all directions. If the floor plan of the building is open enough, the direct air blast will propagate through the building almost unimpeded, ejecting the building's contents through the lee-side windows at high speeds.

Large aerial explosions do not have the same effects as surface bursts. Explosions that occur at high enough altitudes will lay down severe shock and fire damage without the fireball ever contacting the ground. They will not excavate a crater, and therefore raise far less dust than surface bursts with the same yield. Consider a 100-megaton explosion: if the explosion is on the surface, the seismic disturbance is enhanced and the fireball irradiation of the surface, while locally intense, does not reach far from ground zero. This is because the fireball, once it has risen high enough to be seen, say, 75 km away, will have already cooled below the point at which it can ignite fires at a distance. The same explosion occurring at 10 km altitude would sear and shock a much larger area severely enough to cause extreme damage. But an explosion in the rarified air at an altitude of 50 km would not pack much punch: the fireball would expand to huge size, cool down, and slow to subsonic speeds before encountering the dense lower atmosphere. There is in fact an optimum height for airbursts, at which they devastate the maximum amount of area on the surface. Both the Hiroshima and Nagasaki explosions were suboptimum airbursts that caused no cratering and had minimal fallout. The optimum burst height is much too high for cratering to occur. There is also a maximum height for causing any devastation on the ground. Explosions above that height may cause widespread light damage, but even directly below the blast buildings are not demolished. For a 1-megaton explosion that maximum height is about 7 km (4.2 miles); for 10 megatons, about 13 km (8 miles); for 100 megatons, about 29 km (17 miles). In general, the optimum burst height for surface devastation is about 36 percent of the maximum height at which surface destruction of the sub-fireball point can be achieved. For a 1-megaton blast, the optimum height is 2.5 km.

Large explosions heat vast masses of the atmosphere to such high temperatures that nitrogen is partially burned to make toxic nitrogen oxides. The mushroom cloud from a large explosion is always tinged a dirty brown by the presence of nitrogen dioxide gas. Downwind from the explosion, the nitrogen oxides react further with oxygen and water vapor in the atmosphere to make nitric and nitrous acids, which, when washed out of the atmosphere by precipitation, fall as acid rain. The nitrogen oxides are almost as effective as dust at screening out sunlight.

Shallow underground explosions resemble surface bursts in many respects, but the fireball fails to emerge from the ground at greater depths of burial. Our substantial experience with deep underground nuclear explosions, of little relevance to our present interests, has arisen as a natural result of the desire to minimize radioactive fallout from weapons tests.

Several nuclear weapons tests have taken place on or under water. Since roughly three-fourths of the impact events on Earth occur in water, the results of these tests are of great interest to us. An underwater explosion inflates a cavity in the water as the fireball expands. For all cases of interest to us, the explosions are "shallow" in the sense that the fireball will break out into the atmosphere and generate a conical jet of ejected water, steam, and impactor vapor. The cavity will expand to the point at which the diminishing pressure of the expanding, cooling fireball will be overbalanced by the pressure of the water. The "transient cavity" in the water will then collapse, overshoot, raise a high column of water in the center, collapse again, and oscillate several times. Giant waves following the water surface will spread out at speeds far faster than a ship can move. In deep water the height of the wave drops off slowly with distance: airburst shock waves expand in three dimensions, and their strength drops off with the square of the distance from their source. But tidal waves (tsunamis) from a localized source expand in the plane of the ocean's surface. Their height drops off only linearly with distance. When these waves encounter a coastline, they sharpen up their profile due to friction between the ocean bottom and the advancing wave fronts. The wave height grows by about a factor of thirty, so that a thirty-kilometer-long wave with an open-ocean height of thirty meters would crest at a height of about one thousand meters as the wave runs up on a shoreline.

Ocean impacts will rarely possess enough energy to penetrate the ocean completely and crater the ocean floor; neither will they easily raise vast clouds of dust. The main geological evidence for a major ocean impact would lie in tsunami wave deposits. These are usually relatively perishable surface features, such as massive log piles from downed forests, which will rot or weather away in centuries, leaving no clear and unambiguous evidence of the existence of an event that had profound global consequences.

With this body of lore on the (nonnuclear) effects of large nuclear explosions, much of the data on the Meteor Crater and Tunguska events now make sense. The ignition of the trees in the area around the Tunguska airburst is a direct result of heating by thermal radiation from the fireball. The pattern of downed trees, with stripped, singed trunks standing near the center, and all trees felled radially outward farther from ground zero, makes perfect sense as the result of near-vertical impact of the blast wave under the explosion, with more glancing incidence farther out. The fires were apparently rather thoroughly extinguished by the arrival of the air blast wave, a result that fits well with the difficulty of having a firestorm in a sparse, wet boreal forest growing in boggy, sodden ground. From the pattern of damage at the Tunguska site, it is easy to make an estimate of the yield and height of the explosion. Modern understanding of airbursts shows that a ten- to twenty-megaton explosion took place at an altitude of about seven thousand meters (twenty-four thousand feet), close to the optimum burst height for an explosion of this magnitude. The absence of a crater is expected for an explosion at the optimum burst height.

Meteor Crater is easily understood as the result of a surface impact of a body bearing about fifteen megatons of energy. The upthrusting of the strata around the crater rim, the elevated rim itself, and the pattern of ejecta and projectile fragments around the crater all point to an impactor that penetrated the surface and exploded at modest depth. The near-total vaporization of the projectile, the fused quartz grains, and the tiny droplets of iron rain condensed from the fireball are all direct results of very high fireball energy content.

A twenty-megaton explosion near the optimum burst height will devastate an area in excess of two thousand square kilometers; a rough circle with a radius of twenty-five kilometers. This is the size of any major metropolitan area. Such an explosion could utterly destroy New York or San Francisco, London or Paris, Tokyo or Beijing, Rio de Janiero or Mexico City. The death toll from a single such explosion could easily be over 10 million people. And the explosion would leave no crater, no clear and lasting geological evidence of an extraterrestrial cause—only total devastation.

5

THE SPACE AGE:
THE CRATERED PLANETS

Another one full of holes! A lot of these planets are really behind in maintenance.

ANONYMOUS BYSTANDER
Mariner 10 Mercury encounter
Jet Propulsion Laboratory
March 29, 1974

In the spring and summer of 1957 the space race was still only a competition of words. The United States promised an all-civilian space program, delayed to await the debut of the specially designed Vanguard rocket, and ordered the military satellite program to stand down. The Russians, working energetically to build the first intercontinental ballistic missile (ICBM), popped a small satellite atop their behemoth military booster and launched it into orbit on October 4, 1957. *Sputnik 1,* an 84-kilogram test sphere with no scientific instrumentation, was accompanied into orbit by a 4,000-kilogram empty rocket stage—and Americans lamented that the 84 kg was *six times* the weight of the planned Vanguard satellite. Just a month later, the huge Soviet booster launched a 510-kilogram dog-carrying satellite on its second flight. A month later a hastily assembled American Vanguard test vehicle with a 3-kilogram test satellite spectacularly blew up in front of live TV cameras, and national resolve suddenly crystallized. The American military rocket hardware was released from its ban, and the race was on in earnest.

It was tempting to dredge up the hoary analogy of the tortoise and hare; the plodding American tortoise hoping to overtake the Russian hare, already almost out of sight up the road ahead, but possibly prone to taking an occasional nap along the way. . . . But a more apt comparison would be a cageful of American cheetahs looking after an elephant that had a hundred-yard head start. The American program rapidly moved up from tiny, marginal boosters (the Vanguard and the Jupiter C) to second-generation vehicles, the Thor Able, Thor

. . . .

Agena, Thor Delta, and Juno II, based on intermediate-range ballistic missiles (IRBMs). By the end of 1958 the United States was flying its own ICBM, the Atlas, and by late 1959 a version with the Atlas first stage and the Vanguard-derived upper stage stack from the Thor Able, a combination called the Atlas Able, was ready for flight. The Atlas Agena followed just three months later. By the end of 1958, just as the American ICBMs were appearing on the scene, the United States had orbited seven small scientific satellites aggregating 163 kilograms of payload. The Soviets had by then launched three Sputniks aggregating 1,922 kilograms of payload, including one payload with a dog aboard. Strangely, the 10:1 Soviet advantage in throw weight resulted in a 10:1 American advantage in data return. American satellites used highly advanced, minaturized electronics instead of vacuum tubes and were powered by solar cells instead of lead-acid batteries, giving them a vast advantage in performance and useful lifetime in space.

The Americans, beaten into orbit, immediately sought to find something that they could do first. The Soviets, with their vast lead in lifting power, sought to find something new and spectacular that they could do to reaffirm their lead. Both nations chose to turn to deep-space missions, first probes of the Moon, and then visits to the nearest planets, Mars and Venus. In late 1958 the United States launched three Pioneer spacecraft on missions to the Moon. The marginal boosters used for these missions (the Thor Able and Juno II) were not up to the task, and none of the three reached the vicinity of the Moon. On January 2, 1959, the Soviet *Luna 1* spacecraft, launched by an upgraded version of the same booster that launched *Sputniks 1, 2,* and *3,* flew by the Moon at a distance of 6,000 km (the diameter of the Moon is only 3,400 km). While the Americans scrambled desperately to make their much smaller boosters do this ambitious task, a second Luna, launched on September 9, 1959, impacted upon the Moon at high speed, and the third in the series, launched on the second anniversary of *Sputnik 1,* flew by the Moon and radioed back crude photographs of its far side, the part never seen from Earth. The Atlas Able, a third-generation American launcher, but still only half the mass of the Soviet A-class workhorse booster, tried three times in late 1959 and early 1960 to launch flyby missions to the Moon, but without success.

Meanwhile, in October of 1960, the Soviets launched two puzzling "heavy Sputniks" into low orbit around the Earth. Both entered orbit but quickly malfunctioned. No announcement of their purpose was forthcoming from the Soviets. In February of 1961 two more space-craft of this type were placed in closely similar orbits. One of them coasted in orbit for a while and then ignited a rocket engine that accelerated the payload to escape velocity, heading for an encounter with Venus. This spacecraft, named *Venera 1*, suffered a communica-tions failure en route to Venus and was lost. With this clue, it was a simple matter to look back at the two launched in the previous Octo-ber and realize that the same planetary-mission hardware had been launched then, precisely at the best time for flights to Mars. The space race had broken clear of Earth and was spreading throughout the inner solar system.

Using the Atlas Agena B booster and a sophisticated photographic payload with TV relay of pictures back to Earth, the Americans strug-gled to investigate the Moon with its Ranger spacecraft. From late 1961 through early 1964, six Ranger probes were launched toward the Moon without success. The single Soviet lunar probe attempt in this interval, *Luna 4*, a possible landing attempt, missed the Moon by 8,500 km. *Rangers 7, 8,* and *9,* in late 1964 and early 1965, operated perfectly, taking long series of nested photographs as they sped toward impact with the lunar surface. Between them, they returned a total of 17,331 pictures of the Moon. These pictures were of stun-ning quality, showing a detailed record of intense cratering all the way down to sizes of a meter or less. The continuous record from giant basins hundreds of kilometers across to tiny meter-sized craters, and the utter absence of any indication of explosive volcanism, made it absolutely clear that the cratering of the Moon is caused by impacts, not internal volcanic activity. In the absence of absolute ages for specific regions, the rate of bombardment could not be calculated accurately, but assuming that the heavily cratered highlands were some 4 billion years old permitted rough estimates of the impact rate of crater-producing bodies through the inner solar system. Supple-mented by precise age dating of materials returned a decade later by America's manned Apollo missions and by three small Soviet auto-mated landers, the cratering rate could be calculated over a very wide range of impactor sizes. Using information from nuclear test pro-grams on how big a crater can be excavated by explosions of various sizes, the size distribution of the impactors could be reconstructed. Then, allowing for the superior gravitational attraction of the Earth, the rate of infall of cometary and asteroidal materials onto Earth could also be calculated quite accurately.

A similar story was told by missions sent to study Mars. Using their first-generation lunar rocket, the A2e (now, in the era of glasnost, known as the Molniya booster), the Soviets launched probe after probe to the Moon, Mars, and Venus. From 1960 to 1965 the Soviets launched twenty-six known spacecraft into deep space. The result: twenty-six consecutive failures. History has shown that the American space program falls into disgrace upon suffering *two* consecutive failures. Yet the Soviets were committed to solving their problems, and never considered quitting. As a direct result, they were eventually very successful. Of the first eight probes launched toward Mars from Earth, six were Soviet failures and one was an American failure. But the first spectacular success went to the American *Mariner 4* spacecraft, which flew by Mars on July 15, 1965. *Mariner 4* shocked the world with its images of Mars. Generations of Americans, raised on scientific and science-fictional assertions of the similarity of Mars to Earth, expected mild surface conditions and a surface not unlike a desert on Earth. But the pictures sent by *Mariner 4* showed a starkly lunar landscape. Somewhat softened by wind erosion, the surface nonetheless looked more like the lunar highlands than like anything on Earth. Huge craters side by side, superimposed on one another, overlapping in endless layers, covered the ground in the best images. *Mariner 4*, a high-speed flyby, actually returned detailed coverage of only a small percentage of the total surface area of the planet. But there was no doubt: vast areas of the Martian southern hemisphere must be very ancient, heavily cratered terrain, similar in age and cratering history to the lunar highlands.

Another pair of Mariners, launched in early 1969, also flew close flyby missions and photographed regions different from those seen by *Mariner 4*. These probes, *Mariners 6* and *7*, returned another 201 photographs of the Martian surface. Again, all those taken close to the planet, near the time of the best resolution, showed heavily cratered terrain. Three independent looks, and three independent sets of results, pointed to one conclusion: Mars was a dead, Moonlike planet; its atmosphere a sham, a false promise of a kinder, gentler planet that does not exist, the Mars of Edgar Rice Burroughs, canals, and luxuriant vegetation. Forget that cruel illusion: Mars is, and always has been, dead—or so it seemed!

Although sobered in our expectations, with our hopes of exploring an Earthlike planet and searching for life dashed, the people of Earth had already paid for the development and construction of eight additional spacecraft that had not yet been launched. Included in the group were six heavy, complex payloads carried by the Soviet's

D1e (Proton) booster, a vehicle with three times the lift capacity of the already large A2e booster. In what seemed like the last gasp of Mars exploration, these six, plus two American spacecraft, were launched in the May 1971 launch window. First from the blocks at Cape Canaveral was the American *Mariner 8* spacecraft, intended to orbit Mars and map its entire surface from orbit. The second stage of its Atlas Centaur booster failed, and *Mariner 8* crashed into the Atlantic Ocean. Two days later, a Soviet booster carrying two spacecraft, an orbiter and a lander, blasted off from the Tyuratam launch site in Soviet Kazakhstan. The Mars payloads were successfully inserted into low orbit around Earth, but their rocket engine failed to ignite to propel them outward to Mars. Nine days later another Soviet orbiter-lander pair was launched successfully into a Mars-bound trajectory, and nine days after that a third pair of Soviet spacecraft was also sent safely on its way. Finally, on May 30, the American *Mariner 9* spacecraft, another orbital photographic mapper, was sent off. The *Mars 2* and *Mars 3* orbiters and *Mariner 9* all made it safely into orbit around Mars, arriving in the midst of one of the greatest dust storms ever seen on the red planet. The Soviet landers, pre-programmed for entry, fell blind into the howling maelstrom of dust and were destroyed by winds that apparently reached speeds of 500 kilometers per hour (about Mach 0.4). The *Mars 2* and *3* orbiters wasted most of their lifetimes waiting for the dust to subside so that they could see the Martian surface. *Mariner 9*, with a much more capable and flexible computer system, bided its time (some say, by playing solitaire) until the dust cleared, then transmitted over seven thousand superb photographs to Earth.

The results from *Mariner 9*'s mapping survey of Mars were as shocking as the earlier results from *Mariners 4, 6,* and *7.* Simply put, the three earlier flybys had by ill fortune all chanced to get their best views of the half of the Martian surface that is ancient and heavily cratered. *Mariner 9*, with its vastly superior vantage point in orbit around Mars, saw it all. The other half was well worth waiting for: huge volcanoes, vast systems of dry riverbeds, ancient lake and sea basins, evidence of kilometer-thick deposits of ground ice and ancient glaciation. Mars is a two-faced planet. The "southern" half (actually, the dividing line is tilted about thirty degrees relative to the equator) is indeed cratered, Moonlike, and dead. The other half betrays powerful internal processes and the former presence of running water. Some of the surface is very young, and therefore almost free of craters. Some geological activity may even be going on today. And there is a real possibility that, early in its history, the earliest steps of chemical evolution along the way to life may have been taken.

Based on these exciting results, the search for life on Mars gained new life of its own. The Soviet Union launched *Mars 4, 5, 6,* and *7* (two orbiters and two landers on four Proton boosters) in 1973, and NASA launched two Viking Orbiters and two Viking Landers in 1975. But our concern is not with the search for life on Mars: it is with the bombardment history of the planets, and what their impact history tells us about Earth's.

The last of the terrestrial planets to be visited by spacecraft from Earth, and the hardest to reach from here, was Mercury. *Mariner 10,* also known as the Mariner Venus-Mercury flyby (MVM) was launched from Cape Canaveral aboard an Atlas Centaur vehicle in November of 1973. The principal target, Mercury, was so small, so distant, and so close to the Sun in the sky that study of its surface from telescopes on Earth was difficult almost to the point of impossibility. Only a few large, blotchy markings could be discerned by telescopic observers. All available evidence suggested a scorched, airless, waterless surface, but a much closer look was needed to be sure of anything.

Mariner 10 flew by Venus at close range, using the gravitational attraction of the planet to divert the spacecraft trajectory inward toward the Sun, enough to intersect the orbit of Mercury. In its orbit around the Sun with its perihelion at Mercury's orbit, and with an orbital period of two Mercury years (176 Earth days), *Mariner 10* flew by Mercury at high speed on March 29, 1974. One Mercury year later, Mercury was at the same point on its orbit, but *Mariner 10* was at aphelion on the opposite side of the Sun. At *Mariner 10*'s next perihelion passage, Mercury was again in position for a close flyby, which occurred on September 21, 1974. A third flyby was carried out two Mercury years later, on March 16, 1975. All three passages afforded a superb view of Mercury's surface.

Mercury's unusually eccentric orbit carries it in as close as 0.306 AU (perihelion) and takes it as far out as 0.467 AU from the Sun (aphelion). This wide range of distances causes the tidal force of the Sun on Mercury's equatorial bulge to be weaker by a factor of 3.6 at aphelion than it is at perihelion. Tidal forces near perihelion have caused the rotation of Mercury to become locked on to the Sun, so that the angular rate of rotation of Mercury is equal to the angular rate of Mercury's orbital motion around the Sun near perihelion: for a few Earth days near the time of perihelion passage, a particular place on Mercury's equator keeps the Sun almost perfectly overhead.

However, since the rotation rate of Mercury is constant and its rate of travel along its orbit varies with its distance from the Sun, this "resonant" lock cannot be maintained when Mercury is farther from the Sun.

The traditional astronomical opinion before the dawn of the era of modern exploration of the solar system was that Mercury rotated exactly once in a Mercury year. It would therefore always point the same general region toward the Sun, and a portion of the other side would be in perpetual darkness. A "hot pole" and a "cold pole" would coexist at opposite points on the equator. But in reality, Mercury has accommodated itself to a unique arrangement: Mercury rotates on its axis exactly three times for every two Mercury years. After a single Mercury year (from one perihelion passage to the next) the planet rotates exactly one and a half times. Thus a spot on the planet that points at the Sun at the time of the second perihelion passage is *exactly opposite* the point on the equator that was seared at the previous perihelion. After a second Mercury year, the original spot is back at the subsolar point at perihelion. Thus Mercury has two "hot poles" on its equator.

The ramifications for our understanding of Mercury are twofold. First, the three *Mariner 10* flybys at alternate Mercury perihelion passages each found the planet with exactly the same areas illuminated, and exactly the same half of the planet in darkness. Thus only half of the planet could be photographically mapped. Second, the coldest spots on Mercury would now be relegated to low, shadowed basins near the poles, requiring only modest retargeting of the spacecraft trajectory to see both polar regions. A permanently shadowed cold basin cannot, however, be photographed, because it is hidden in perpetual darkness.

The photographic mapping of the accessible half of Mercury was a great success. Almost all of the surface was covered with layer upon layer of huge craters. The scene was reminiscent of the lunar highlands. A few very large impact basins with lava-flooded floors and lower crater counts were seen to penetrate the thick, heavily cratered crust. The evidence for internal geological evidence was there, but the general conclusion was clear: Mercury's surface is overwhelmingly shaped by external forces. Impacts are hundreds of times more effective in shaping the surface than internal volcanic and tectonic processes. The surface of Mercury is a diary that eloquently attests to the fierce bombardment of all the terrestrial planets, but it is unfortunately a diary without dates. That cratering has been an immensely powerful process is obvious, but the rate at which large craters are presently forming on Mercury can only be estimated. Combined with the context of the Moon and Mars, Mercury's fate tells us that Earth lies well within the region of the solar system that was heavily bombarded. However, Earth's nearest planetary neighbor, Venus, remained an enigma. Given that Venus has a dense, perpetual cloud cover, it was not possible for *Mariner 10* or any other photographic flyby or orbiter mission to search for cratering on its surface. Thus the questions inevitably arose, "What happened to Earth? Where is the evidence?" Meteor Crater was merely the first documented impact crater on Earth. Its authentication opened the door to hundreds of others.

In the 1940s and 1950s, using the exploratory tool of aerial photography, several large impact craters were discovered on the Canadian Shield, one of the oldest and stablest portions of Earth's crust. Three small craters near the Quebec-Labrador border were studied in 1954, and a 3-km crater was identified near the town of Brent in Ontario. In 1955 the resources of the Canadian Air Photo Library were placed at the disposal of crater-hunters, and more discoveries quickly followed. The 12-km Deep Bay crater in Saskatchewan was discovered from the air. The striking pair of Clearwater Lakes, about 33 km and 22 km, respectively, soon followed, along with the 65-km Manicouagan crater in Quebec. All of these craters were well established by 1965.

Verification of the meteoritic origin of these and many other craters was assisted by the high-pressure synthesis and discovery, in 1953, by chemist Loring Coes, of a mineral identical in composition with normal quartz, but much denser. The naturally occurring mineral of this structure, appropriately called coesite, requires such extreme conditions for formation that it is found in nature only where silica-rich rocks have experienced the extremely violent compression of a meteorite impact. Once coesite had been found in impact craters (in 1960), it joined the several other indicators of impact such as meteoritic iron, magnetite spherules, shatter cones, and impactite glass to provide "ground truth" verification of impact craters discovered from aerial photography.

One of the greatest gifts space technology has given the human race is the ability to look at Earth from above, with a truly global and holistic perspective. In keeping with the simian heritage of our mental and social apparati, it was only natural to seek both inspiration and practical advantage from this new dimension of exploration. Even years before the first artificial Earth satellites were launched, it was already clear that photography of Earth from space would have a wide range of profound applications in areas as diverse as weather prediction, ICBM targeting, crop and pollution monitoring, mission planning for strategic bombers, Earth resources surveys, and arms control monitoring and verification. In 1958, when *Sputniks 1, 2,* and *3* were newly launched, President Charles de Gaulle of France confronted Nikita Khrushchev with press reports that the Soviet satellites carried espionage cameras to photograph military installations in Western Europe. This was an especially sensitive matter, since official Soviet political doctrine asserted that total war with the capitalist nations was inevitable, and official Soviet military doctrine argued that a massive surprise attack against all possible enemy strategic weapons bases was the the only way to assure victory. Khrushchev responded, possibly with some astonishment, and surely with the wounded pride he affected so well, that there were no such cameras on Soviet spacecraft. De Gaulle, not so easily appeased, and little inclined to trust Soviet assurances, responded "So you say!" In historical perspective, we know that Chairman Khrushchev was surely right. Soviet electronics technology would not be ready for TV-type high-resolution imaging spacecraft for more than a decade. Photographic film cameras on satellites would be of no use until the means of recovering film capsules from orbit could be developed. The latter effort got under way with a series of launches beginning in May 1960. It is generally believed that the earliest Soviet military use of photographic imaging spy satellites was in 1962. The *Kosmos 4* and *7* flights, ostensibly preparations for manned missions, may have carried out experiments along these lines, and the secret military missions of *Kosmos 9, 10,* and *12* later in the year all involved recovery of hardware from orbit. De Gaulle was quite correct in principle, but about three years ahead of his time.

The first photographic reconnaissance by American spacecraft actually occurred slightly earlier than that. The air force program to develop recoverable film capsules, Discoverer, began flights in early 1959. It was not until the film capsule from *Discoverer 13* was successfully recovered from orbit, on August 11, 1960, that the first manmade object was safely returned from orbit. The space race was then at its hottest: the first Soviet recovery of a satellite occurred only nine days later. It was a year before the operation of the Discoverer recovery system was reasonably reliable. Indeed, it was in August 1961 that *Discoverer 29* returned superb photographs revealing the Soviet operational ICBM base near Plesetsk, which the Soviets referred to internally as the "Northern Cosmodrome." For more than twenty years thereafter the Soviets continued to deny to the world that such a base even existed.

Civilian imaging of Earth was limited by the desire to avoid Soviet accusations of espionage. The Tiros weather satellites, with an early TV system for transmitting cloud-cover images to Earth, became operational in 1960. The optics of these satellites were intentionally degraded to a resolution of one kilometer, making the detection of even the largest weapons system impossible. At this resolution, Los Angeles looks like a gray blur. (Recent visitors report that, in reality, it more closely resembles a *brown* blur.) With the hardware development of the Soviet photographic reconnaissance satellites already well under way, Khrushchev attacked the Tiros system as a provocative infringement of Soviet sovereignty (this, remember, was a system that could not even verify the existence of Moscow) that would have to be answered by superior Soviet technology. The secret American Discoverer satellites, not mentioned by Khrushchev, flying as part of the Corona program, were meanwhile demonstrating that they could resolve one-foot objects on the ground.

Civilian imaging of Earth's surface with useful resolution was begun during the Gemini program using a handheld Hasselblad camera operated by the Gemini astronauts in 1965 and 1966. NASA followed with a series of missions devoted to Earth resources mapping and monitoring, which included photography with about thirty meters resolution in several different wavelength bands selected for their usefulness in discriminating between different materials on the surface. Such studies began with the Landsat program in 1972 through 1984, and were continued by the Soviet Priroda program and the European Space Agency's Spot spacecraft. Judicious combination of satellite and aircraft photographic mapping, fieldwork, and laboratory study of materials collected in the field has now extended the list of authenticated impact craters to over two hundred. Many new craters are added to this list each year. Thus, despite the intense erosion of Earth's surface and the recycling of old crust, despite the high probability that impactors will land in the oceans or in soft sediment from which all trace of a crater will quickly erode away, Earth has demonstrably received a rain of massive, highly energetic impactors. So far the biggest impact crater is the 200-km-plus diameter Chicxulub crater on the north shore of the Yucatán Peninsula, a 65-million-year-old impact scar well filled by more recent sediment. If this impact had occurred in deep water its traces on the ocean floor might be extremely hard to recognize. Furthermore, ocean-floor rocks are recycled on time scales of tens of millions of years. Most of the ocean floor on Earth has been formed since the time of the Chicxulub impactor, and most of the ocean floor then present has since been subducted into the mantle and destroyed.

Combining all we know about the cratering of Earth, the Moon, Mars, and Mercury, it is possible to make consistent estimates of the flux of cometary and asteroidal bodies in the inner solar system: on Earth, hundred-megaton impacts occur on the average once every fifteen hundred years, and one-thousand-megaton (one gigaton) blasts occur at an average rate of one per twelve thousand years. Teraton impacts (a million megatons; one thousand gigatons) happen about once every eight hundred thousand years. Globally devastating impacts in the billion megaton (one petaton) range happen about every 100 million years.

Explosions smaller than one hundred megatons are much more common, but involve bodies so small (one hundred meters diameter and less) that photographic search techniques cannot detect them. The data on craters made by these smaller bodies get progressively less reliable toward smaller sizes because of the problem of distinguishing impact craters from secondary craters produced by debris ejected from large impacts. Thus a very important part of the bombardment flux on Earth is due to bodies so small that extremely sensitive search techniques are required to find them. Their effects, which are not felt globally, may nonetheless be very severe locally. The problem they present is addressed in the next chapter.

6

NEAR-EARTH OBJECTS

In June 1968, the asteroid Icarus, a dark boulder a mile or so in diameter, will pass by Earth at a relative velocity of about 100,000 fps and a distance of 4 million miles. The orbit of Icarus has been quite accurately established, and the chance of it approaching much closer is nil. However, 4 million miles is an uncomfortably small miss distance in the scale of the Solar System.

Announcement posted at M.I.T. in January 1967

The first century of asteroid research, commencing with the discovery of Ceres in 1801, found 463 asteroids. All but one of these orbit in the asteroid belt, safely beyond the orbit of Mars. The sole exception, Eros, found in 1898, closely approaches Earth. Searches for asteroids in the twentieth century have raised the total of known asteroids with accurately determined orbits to over 6,000, with several hundred more being added each year. Studies of the spectrum of sunlight reflected from these asteroids have given us useful information on the minerals present in them. Hundreds of asteroid spectra have been compared with laboratory reflection spectra of meteorites and pure mineral samples. About two dozen distinct types of asteroids have been found by this technique, of which some can be confidently linked to particular classes of meteorites. Most of the other asteroid spectra clearly show that those bodies, although not identical to familiar meteorite classes, are nonetheless made of the same minerals we see in meteorites.

The asteroid belt is broadly zoned into bands of different classes of asteroids. Most of the inner half of the belt is composed of stony (S-type) asteroids containing the silicate minerals and metal found in ordinary chondrites and stony-iron meteorites. The outer half of the belt is dominated by the C-type asteroids, very dark materials that closely resemble carbonaceous meteorites. The outermost asteroids belong to the P and D types, which seem to be made of the same ingredients as the C types but in somewhat different, more carbon-rich, proportions. The innermost asteroids include a number of bodies with the reflection spectra of metallic iron-nickel meteorites (the

M-type asteroids), and others that look like laboratory samples of various melted and recrystallized silicate meteorites, the achondrites. But the belt is far away. Aside from a few science fiction writers and an even smaller number of solar system astronomers, few people knew anything about asteroids before 1968.

The popular perception of the remoteness and unimportance of asteroids was shattered by a visitor from outer space. In May and June of 1968 the media resounded with news of a close flyby of the asteroid Icarus, then tearing by only 6 million kilometers from Earth at a blazing speed of thirty kilometers (eighteen miles) per second. The orbit of Icarus was well enough known that it was clear no collision would occur; nonetheless, the orbit of this deadly projectile clearly makes Icarus a recurrent long-term threat to Earth. Anticipation of Icarus's June 14, 1968, close pass stimulated a study during the spring semester of 1967 in the Advanced Space Systems Engineering course at M.I.T. The assigned problem: Suppose the projected orbit of Icarus showed that a collision with Earth would occur in June. What could be done about it? Would it be best to try to smash the asteroid into rubble with a huge nuclear warhead, or to try to divert it just sufficiently to change the timing of its crossing of Earth's orbit, so as to assure a miss? The twenty-one students who enrolled in the course formed teams devoted to asteroid science, mission design, the effects of very large (one-hundred-megaton) nuclear explosions, launch systems, spacecraft design, guidance and control, and so on. The students were briefed and advised by an impressive assortment of experts, including Dr. Fred Whipple, then director of the Smithsonian Astrophysical Observatory at Harvard, and an M.I.T. professor of aeronautics and astronautics named James A. Fletcher, who was later to serve two separate terms as administrator of NASA.

At the end of the course the students staged a three-hour final presentation for the M.I.T. community. Among the guests were a host of reporters, and stories about the study were carried by at least thirty newspapers, many of which sensationalized and seriously distorted the circumstances of Icarus's pass. Among the more responsible publications were *The Boston Globe*, where Project Icarus made the front page, and the June 16 science section of *Time* magazine. Thus many Americans for the first time became aware both of the probability of an asteroid impacting the Earth and the possibility that something could be done about it.

But to astronomers, the close flyby of Icarus was nothing new.

Indeed, Icarus had been discovered by astronomer Walter Baade during an equally close approach in 1949. Icarus was then not the first, but the thirteenth asteroid discovered in an orbit that takes it close to Earth. Eros, the first of these Near-Earth Asteroids (NEAs) to be discovered, follows an orbit that grazes Earth's orbit when Eros is at its point of closest approach to the Sun, but the two orbits do not presently cross and a collision is not now possible. All of the first six NEAs to be discovered, through the asteroid Amor in 1932, were members of the class of Earth-approaching bodies that do not now cross Earth's orbit. Almost all of the members of this group of presently nonthreatening NEAs, called the Amor family, are in orbits that cross the orbit of Mars. They can therefore be perturbed by gravitational interactions with Mars into orbits that directly threaten Earth.

The next six NEAs to be discovered (starting with Apollo in 1932, and running through 1949) were all in orbits that actually cross Earth's orbit. Several of them cross Venus and Mars as well as Earth. The unlucky thirteenth NEA discovered, our friend the Apollo asteroid Icarus, crosses Mercury, Venus, and Earth: at perihelion, Icarus approaches to an astonishing 0.187 astronomical units from the Sun, less than half Mercury's distance from the central fires of the solar system. These asteroids with Earth-crossing orbits and orbital periods greater than one Earth year constitute the Apollo family.

Over the following few years, several more Apollo and Amor asteroids were discovered. Late in 1954 a faint, half-kilometer asteroid named 1954 XA was discovered in an Earth-crossing orbit with a period less than one Earth year. This asteroid has since been lost, but another well-documented asteroid of this family, Aten, was discovered in 1976. These bodies that spend most of their orbits inside Earth's orbit are termed the Aten family. The three groups of near-Earth asteroids are collectively termed the AAA asteroids. Their known and projected populations are given in Table 1.

It is reasonable to wonder whether there might be other asteroids with orbital periods less than one year that do not presently cross Earth's orbit but graze it from within. Since we discover asteroidal bodies by looking for them in the late-night sky, however, we would never see any of these bodies, even if they were numerous. They remain hidden from our scrutiny by the glare of the daytime sky. But, despite their undetectability, such bodies might at any time pass close to Venus and be perturbed by Venus into an Earth-crossing orbit.

. . . .

The pace of AAA discoveries accelerated as their presence became more widely known. Eight new NEAs were discovered in the 1950s and seven more in the 1960s, all as serendipitous results of other observations. Then, in 1960, in a dedicated asteroid search called the Palomar-Leiden survey, a young Dutch astronomer, Tom Gehrels, turned up four new NEAs. The 1970s brought twenty-four more discoveries. By 1982 the discovery rate had reached a remarkable rate of ten per year, thanks largely to the concerted efforts of a team of astronomers led by Eugene Shoemaker and Eleanor (Glo) Helin. Encouraged by their success, other observers, including Rob McNaught in Australia, established systematic photographic search programs to discover near-Earth bodies. The traditional way of discovering asteroids relies upon using a small, relatively fast telescope to produce a number of photographic plates of the dark (moonless) sky in the direction opposite the Sun. The observer patiently monitors the aiming of the camera all night, photographing and then rephotographing a limited portion of the sky, and then hastens at the crack of dawn to develop the plates taken during the night and scan them, using a device called a blink comparator, to see whether any bodies in the field covered by the plates have moved during the night. This exhausting and tedious work occupies the two weeks of "dark time" bracketing the date of the new Moon. By such means, an average of eight new NEAs were discovered each year from 1983 through 1988.

These photographic searches have limited ability to find small, faint, fast-moving targets because the image smear caused by their rapid angular motion makes them too faint to register on photographic plates. Generally, it is very difficult to find asteroids smaller than about 1 kilometer in diameter using photographic techniques. The smallest NEAs ever found photographically are 200–300 meters in diameter. This means that asteroids up to about 250 meters in diameter can generally evade detection indefinitely. This sounds like a small body, but its energy content is actually enormous: a 250-meter asteroidal rock traveling at orbital speed typically carries a kinetic energy of about thirty thousand megatons of TNT. This is more than the total explosive energy of all the nuclear weapons on Earth. Thus the best photographic detection techniques available in the 1980s *could not even find asteroids with the potential to inflict damage worse than World War III.* Fortunately, in 1989, a new, ultrasensitive high-tech search program entered upon the scene.

The Spacewatch program, conceived and directed by Tom Gehrels in the Lunar and Planetary Laboratory of the University of Arizona, was designed to remove the bottleneck of photographic imaging, developing, and plate scanning from the NEA discovery process. Spacewatch makes extensive use of modern electronic detectors and computers to automate much of the discovery process. The 90-cm (36″) Spacewatch telescope on Kitt Peak, Arizona, is used for half of the nights of each month (during "dark time" when the Moon is not near full) to scan the sky with a CCD (charge-coupled device) detector array, a sensitive solid-state electronic detector related to those in portable video cameras. The Spacewatch CCD detector is of a type originally developed for military satellite reconnaissance use. It consists of a square 2048x2048 array of tiny light-sensitive detector spots, with associated electronics to "read out" the light intensities measured by these detector spots in sequence. These intensities can be used directly by a video monitor to build up an image with 2048x2048 "picture elements," generally called pixels in the trade. In addition, the digital intensity information from these pixels can be processed by a computer to build up an exhaustive catalog of all the objects, including stars, galaxies, belt asteroids, comets, and NEAs, in the image. The computer-controlled telescope stores these images on magnetic media. After the area chosen for that night's survey has been covered completely, the computer directs the telescope to resurvey the area covered to date. The computer instantly compares each new image to the view of the same area recorded earlier in the night. Almost everything recorded in the two images will be in the same place, and the computer checks off the objects seen in the second image against the catalog made from the first image of the same area. Any feature that appears in only one of the images is noted (these are mostly defects, electronic interference, or cosmic ray strikes) and all those that have moved are flagged and brought to the attention of the operator. Some images are taken to refine the positions and orbits of bodies discovered on previous nights or by other search programs. Finally, a third round of images is collected late in the night, to verify that the bodies that seem to move between the first and second images continue to do so in a consistent manner. Almost all of these moving objects with systematic motion are rather slow-moving belt asteroids, out between Mars and Jupiter.

Spacewatch discovers hundreds of new kilometer-sized belt asteroids with every observing run. These observations are reported to a central astronomical clearinghouse for asteroid and comet data at the

. . . .

Smithsonian Astrophysical Observatory in Cambridge, Massachusetts, where Brian Marsden and his staff maintain a giant database of observations and calculate orbits for newly discovered bodies. Spacewatch itself, however, does not follow up on these belt observations: its limited observing time is dedicated to the discovery and follow-up of more interesting—and dangerous—objects.

A small proportion of the bodies discovered by Spacewatch are moving too fast (or too slowly) to be belt asteroids. The fast-moving ones are usually small, faint, nearby asteroids dashing by Earth at relatively close range: their nearness gives them very high angular rates of passage across the sky. These fast-moving objects include the near-Earth asteroids and nearby comets.

Spacewatch became operational in the fall of 1990 and immediately began setting records. Spacewatch's fourth discovery, 1990 UN, was the smallest asteroid yet discovered, a mere one hundred meters in diameter. Just three months later, Spacewatch found 1991 BA, a tiny body only nine meters in diameter (about thirty feet!), the smallest and closest astronomical body ever found. The following fall, Spacewatch discovered 1991 VG, a tiny eight-meter object with an orbit astonishingly similar to Earth's. Another ten-meter body, unglamorously named 1994 ES1, was discovered in March of 1994 at half the Moon's distance from Earth. Several other asteroids with exceptionally Earthlike orbits have since been added to the list, and several more bodies smaller than ten meters have been found. These bodies in Earthlike orbits are not only the solar system bodies most accessible to Earth, but also the most likely to be drawn into collision with the planet by Earth's gravity: they approach at such low speeds that they find Earth irresistibly attractive.

Spacewatch has found that the number of small asteroids is larger than a simple extrapolation of the number of larger bodies would suggest. At sizes of one hundred meters, the Spacewatch figures match the expected number well. Spacewatch finds about twice as many bodies at thirty meters diameter as the linear extrapolation predicts. At ten meters, the excess is about a factor of ten. At three meters, Spacewatch data project an excess of a factor of thirty or so. Objects with diameters of one to five meters are usually too small for Spacewatch, but meter-sized objects are occasionally observable as fireballs when they enter the atmosphere. Thus any extrapolations of Spacewatch data to smaller sizes can be checked by comparing them to the observed frequency of bright fireballs.

Perhaps the most important contribution of Spacewatch has been a dramatic acceleration of the NEA discovery rate. Nine NEAs were found in 1988, and fifteen in 1989 (one by Spacewatch). In 1990 twenty-seven NEAs were discovered (six by Spacewatch). In 1993 Spacewatch was responsible for 60 percent of all the NEAs found.

The other major contribution of Spacewatch is the discovery that large numbers of these NEAs are following orbits of very low eccentricity, about 0.1 compared to 0.5 for a typical Apollo asteroid (and 0 for a perfect circle). Further, they have orbital periods very close to one year. In short, these bodies are closely following Earth's orbit around the Sun. Tom Gehrels has informally named these bodies in Earthlike orbits the Arjuna family, after the chief protagonist of the Bhagavad Gita, although they are all in fact members of the well-established Apollo orbital family. At first it seemed plausible that these bodies might be meter-sized rocks hurled off the Moon by comet and asteroid impacts, which received just enough impulse to escape from the Earth-Moon system and go into orbit directly around the Sun. But this interpretation became much less plausible when it was found that some of these bodies are larger than one hundred meters in diameter. It simply does not seem possible to blast so large a body off the Moon without crushing it to rubble.

Since these bodies follow Earth's orbit rather closely, they approach Earth very slowly. Because of their low relative velocity, they are subjected to Earth's gravitational forces for unusually long periods of time, and find it difficult to resist Earth's attraction. The probability that such a body will be deflected into an impact trajectory is much higher for the Arjunas than for other Apollos. The Arjunas that are also blessed with low orbital inclinations are especially susceptible to Earth's blandishments. Their low orbital inclination also compels them to be very close to Earth, not far out of the plane of the solar system, at the times when they pass by. Because of these orbital properties, some of these bodies have very short expected orbital lifetimes before colliding with Earth. Many Apollos have expected lifetimes of tens of millions of years to over 100 million years. But some Arjunas are so vulnerable to Earth's gravity that their expected lifetimes are a thousand times shorter. Arjunas must make up a surprisingly large fraction of the bodies striking Earth. Since they are both numerous and short-lived, there must be a powerful mechanism for replenishing them as quickly as they are destroyed. Further, the Arjunas can enter Earth's atmosphere at speeds little larger than the legal minimum—Earth's escape velocity. This is significant because, like the Tunguska impactor, the ability of a body to

. . . .

survive in the atmosphere depends on whether its crushing strength exceeds the aerodynamic pressure generated by its passage through the atmosphere. The aerodynamic "ram" pressure is proportional to the square of the velocity of the body: slow bodies are much better able to penetrate deeply into the atmosphere than fast ones.

Another implication of the Earthlike orbits of the Arjuna asteroids is that they are the easiest solar system bodies to reach from Earth. Many are so accessible that a given rocket booster could land much more payload on the asteroid than it could land on the Moon. Such a mission is by no means beyond present technology. Indeed, as far back as the 1970s the Soviets successfully executed three separate automated missions (*Lunas 16, 20,* and *24*) to return samples from the surface of the Moon. However, landing on and returning material from a nearby asteroid is vastly easier than the comparable maneuvers on the Moon because the asteroid's gravity is so feeble. A 100-meter Arjuna has an escape velocity under 0.2 meters per second, compared to 2,400 meters per second for the Moon.

Such ease of access means that very large masses of asteroid materials can be returned to the vicinity of Earth. A water-bearing asteroid in a nearby orbit could provide large masses of propellant for use in the exploration and exploitation of the solar system. Some transportation system studies have shown that each load of equipment dispatched from a space station to a favorable nearby asteroid could bring back one hundred times its own mass of precious water, structural iron alloys, and rare and strategic metals. Meteorites tell us that the *average* near-Earth asteroid has a higher concentration of precious metals, such as platinum, than the richest known ore bodies on Earth. *Those asteroids that are most accessible to Earth for economic exploitation are usually the very same asteroids that have the highest probability of collision with Earth.*

Some of these small bodies have actually been seen to come very close to Earth. In May of 1993 Spacewatch discovered a faint body that moved an incredible five degrees of arc in the one-hour interval between the first and second round of images. Follow-up quickly revealed that the body was a tiny asteroidal fragment about 3 meters in diameter in an orbit that took it deep inside the Earth-Moon system. Indeed, at the time of discovery, the body was actually retreating from Earth at high speed: just an hour earlier it had flown by Earth at a distance of about 150,000 km, less than half the distance of the Moon. This tiny rock carries enough kinetic energy to produce an explosion equivalent to several thousand tons of high explosives.

It (barely) missed us this time, but it will be back. You can count on it.

Studies of the spectra of about fifty NEAs have given us some valuable clues to their nature and origin. They are very diverse in composition, showing that they sample the belt population very widely. Fully a quarter of the NEAs studied to date are very black and carbonaceous (C type). Interestingly, there are strong selection effects that discriminate against the discovery of the darkest asteroids: we search for them using visible light, which they absorb with great effectiveness. It has been estimated that the real proportion of C-type asteroids must be about twice as large as the 25 percent we find among the ones whose spectra have been studied. There may be systematic changes of the spectra with size, but so few of the small asteroids have been studied that we do not yet know if there are such trends. C-type asteroids of any size are very valuable resources: roughly 40 percent of the mass of a CI chondrite consists of readily extractable volatile elements, notably oxygen, carbon, hydrogen, nitrogen, sulfur, and chlorine. These elements are of great value in making life-support materials, propellants, and industrial chemical reagents. About 30 percent of the residue after extraction of these volatiles is a metallic alloy of iron, nickel, cobalt, and platinum-group metals, a material of interest not only as a source of structural metals, but also precious and strategic metals for return to Earth.

Spacewatch has found, and will continue to find, large numbers of bodies smaller than 10 meters in diameter, too small and faint for spectral studies. The number of such bodies in nearby space is so large that developing a complete catalog of their orbits is a gigantic, almost impossible, task. We presently track and maintain accurate orbital data for over two hundred NEAs and several hundred comet nuclei larger than 1 kilometer in diameter. Analysis of the degree of coverage of the sky by asteroid searches and of the rate of accidental rediscovery of already-known NEAs both suggest that there are about two thousand kilometer-sized NEAs in near-Earth orbits. The search for kilometer-sized NEAs is thus only about 10 percent complete at present. Indeed, 5-km-diameter Apollo asteroids were discovered in 1991 and 1992. At present discovery rates of about twenty-five kilometer-sized NEAs per year, it would take us another seventy years to develop a reasonably complete catalog. However, Spacewatch has clearly demonstrated that there is an inexpensive way to accelerate the discovery rate severalfold.

The challenge presented by one-hundred-meter NEAs is relatively discouraging. The number of NEAs down to this size is projected to be between five hundred thousand and six hundred thousand. Thus the process of search, discovery, orbit determination, and tracking for these smaller objects is hundreds of times more onerous than for kilometer-sized objects. The expected number of meter-sized bodies is vastly larger again: several billion of them probably orbit in near-Earth space. Finding and cataloging all of them is not a practical goal. Any sensitive asteroid search will discover many of these tiny bodies, and we will learn much about their abundances, orbits, and physical properties—but we will not find, or even search for, all of them.

For objects with diameters less than a few meters, even a Space-watch-type system is not sensitive enough to detect them reliably as they fly by Earth. But these bodies are so numerous that there is another way to detect their presence: they appear as fireballs when they strike Earth's atmosphere. Most of these fireballs burn up or explode in the atmosphere and vanish without a trace. Some of the strongest survive long enough to drop meteorites on the ground. But *all* meter-sized impactors put on spectacular fireworks displays. Many are much brighter than the full Moon, and some of the largest even rival the Sun in brightness. Clearly, the best way to count the population of meter-sized bodies is to observe their spectacular entry into Earth's atmosphere over a wide area, and for a long period of time. Such events are rare enough so that professional astronomers seldom observe them, but common enough so that thousands of people see them every year.

Fireballs display a wide range of breakup behavior in the atmosphere. Some are rather strong, with crushing strengths of 10 to 100 atmospheres (150 to 1,500 pounds per square inch). These are probably pieces of moderately strong asteroidal stone, similar to ordinary chondrites. Some crush at aerodynamic pressures of 1 to 10 atmospheres. These appear to be pieces of carbonaceous asteroids. Most other fireballs are extremely fluffy and weak, with crushing strengths of 0.01 to 1 atmospheres. These generally have cometary orbits.

From time to time a brilliant fireball appears in the sky, often accompanied by a hissing or buzzing sound. The fireball then explodes, and many seconds or even several minutes later a loud sonic boom is heard widely on the ground. In my readings on this subject I have found a hundred examples of fireballs brighter than the Moon; hundreds of cases where powerful sonic booms have been laid down across great areas of countryside; many dozens of cases of powerful aerial explosions over major cities, usually (especially at night) associated with reports of brilliant fireballs. Many of these larger fireballs leave trails of dust and smoke that persist for many minutes. A small minority drop showers of recoverable meteorites on the ground, but most are utterly disrupted in their final explosion.

Rarer are stories of near misses of Earth by cosmic projectiles. A meteor was seen on July 7, 1872, to follow a near-horizontal trajectory. It appeared to be rising as it faded from view. According to a report published in the journal *Nature* by W. F. Denning, on May 22, 1889, a brilliant, slow-moving, near-horizontal meteor was widely observed over south-central England. It was first seen from Bristol, Reading, and Clifton, and was tracked by many observers over a wide area. The body was brightly luminous over a path of about 400 kilometers, which it traversed in about sixteen seconds, for an estimated speed of 25 kilometers per second. It reached a minimum altitude of about 80 kilometers at a point some 10 km east of Oxford, and was apparently gaining altitude as it faded from view in the south.

On October 27, 1890, observers at Cape Town, South Africa, saw an extraordinary comet with a tail about a half a degree wide and about ninety degrees long. The comet was visible only from 7:45 to 8:32 P.M., during which time it traversed about one hundred degrees of arc. Supposing this was a typical small comet, traveling at about forty kilometers per second relative to Earth, then its observed angular rate of two degrees per minute implies that the comet must have passed within eighty thousand kilometers of Earth, about a fifth of the distance of the Moon. This is a close miss indeed!

Such observations are rare, and the phenomena associated with these close flybys of Earth are elusive. A few minutes of inattention, a thin layer of clouds, a flyby at extreme southern latitudes where observers are sparse, or a close approach to the dayside of Earth, and the event is missed altogether. As it happens, another small, fast-moving comet was seen by observers taking a break at the European Southern Observatory in March 1992. A comet with a clearly visible nucleus of about the first magnitude (as bright as Vega) was seen to dash through twenty degrees of arc in only three minutes. The comet had a nearly circular coma about two degrees in diameter but no obvious tail. Again taking the most probable flyby speed as 40 kilometers per second, typical of long-period comets, this comet must have flown by at a distance of about 20,000 km. Remembering that the diameter of Earth is about 13,000 km, this is very close indeed. Now, if the reflectivity of the comet nucleus were similar to that of the

Moon (which reflects about 7 percent of the sunlight which strikes it), we can calculate that a first-magnitude body at a distance of 20,000 kilometers would have to be about 170 meters in radius. If we use instead the reflectivity of the nucleus of Halley's comet or of a black, carbonaceous asteroid (about 2.3 percent), then the radius of the solid nucleus would be close to 300 meters. A body this size, with the density of a mixture of ice and rock (about two grams per cubic centimeter), traveling at 40 kilometers per second, would have a kinetic energy content as high as fifteen thousand megatons of TNT.

On February 9, 1913, a brilliant fireball appeared over Regina, Saskatchewan, heading eastward. It blazed its way across the heavens over Winnipeg and Toronto and other cities of central Canada and the northeastern United States, in full sight of tens of thousands of observers. Leaving a glowing trail over one thousand kilometers long, it broke into several large pieces as it progressed. Traveling at a speed of about ten kilometers per second, and maintaining an altitude of about fifty kilometers, it passed over New York and headed out over the Atlantic. About two minutes later a similar fireball "procession" was observed in Bermuda, traveling in very nearly the same direction. If all these observations refer to the same body, it flew for four thousand kilometers within the atmosphere, burning and disintegrating for over six minutes. Its ultimate fate is unknown, although the speed suggests that it was traveling too slowly to escape from Earth. It may have entered a temporary elongated orbit around Earth, or it may have burned up over the South Atlantic after overflying Bermuda.

On August 10, 1972, campers at Zion National Park in southern Utah were startled to see a bright daytime fireball appear overhead. The fireball headed almost due north, passing over Salt Lake City and Pocatello, Idaho. Hundreds of astonished observers, including one with a movie camera, saw the fireball from Grand Teton and Yellowstone parks. The fireball passed on over Helena, Montana, and crossed the Canadian border near Lethbridge, Alberta. Still traveling above escape velocity, it slowly rose and grew fainter over Alberta, until it was no longer visible. Evidently it escaped from Earth into an orbit around the Sun that will repeatedly bring it back to precise intersection with Earth's orbit. And one day, Earth will be in the way again. . . .

One of the observers of this fireball happened to be astronomer

L. G. Jacchia. His analysis of eyewitness reports shows that the fireball was bright enough to be easily visible in the daytime along a path 1,500 kilometers long. It approached Earth with a relative speed of 10.1 kilometers per second, and was accelerated to 15.0 kilometers per second by Earth's gravity as it fell toward the top of the atmosphere. Its point of closest approach to Earth was at an altitude of about 58 kilometers over southern Montana. Sonic booms were heard there but were not reported elsewhere along the fireball's ground track. He estimates the maximum brightness of the fireball as −15 to −19 magnitudes, about ten to four hundred times as bright as the full Moon. The body had a diameter of 15 to 80 meters and a mass of at least several thousand metric tons, and possibly as high as a million metric tons. It passed within 6,430 kilometers of the center of Earth. If it had instead passed only 6,410 kilometers from Earth's center, it would almost certainly have been decelerated enough to prevent it from escaping Earth's gravity. It would then have exploded or impacted somewhere in the populated strip of land stretching from Provo, Utah, through Salt Lake City, Ogden, Pocatello, and Idaho Falls. The explosive power would have probably been between one and twenty kilotons of TNT. Worse yet, a multikiloton explosion of mysterious origin in an American city in 1972 might very well have triggered a nuclear war.

Several photographic networks have been established for the purpose of tracking fireballs, calculating their original orbits around the Sun, and locating the landing points of meteorites. One such system, the Prairie Network, operated in the central United States for several years, observed numerous bright fireballs, and led directly to the recovery of one meteorite. A small network in western Canada operated with similar success. In Europe, two photographic tracking stations were set up in western Czechoslovakia under the leadership of Zdenek Ceplecha in the 1960s. In recent years, a number of other stations of the Czech type have been set up in other European countries to form the European Fireball Network. The earliest two-station version of the network observed the fall of the Přibram (Czechoslovakia) meteorite, and several other meteorite-dropping fireballs have been observed.

The European Network has also reported tracking two Earth-grazing fireballs. A fireball of −12 magnitude (weighing in at roughly one hundred kilograms) was tracked over central Europe on February 3, 1985, on a trajectory that took it back out to escape from Earth.

. . . .

Another fireball, seen on October 13, 1990, also skipped out of Earth's atmosphere. Its closest approach to the ground was an altitude of ninety-eight kilometers over the Czech/Polish border.

Not all close calls are benign. At rare intervals, large bodies enter the atmosphere almost horizontally and actually survive deep into the atmosphere, sometimes even all the way to the ground. A 1965 study of the Campo del Cielo iron meteorites and associated small craters in Argentina showed that they fell along a narrow eighteen-kilometer-long ground track. Individual meteorites of different sizes were well sorted in sequence of mass along the track, evidently by aerodynamic (drag) forces. Disruption of the parent body occurred at an altitude of several kilometers. Radiocarbon dating of charcoal from one of the craters suggests that the event occurred well within the time of human occupancy of South America, about 2900 B.C.

Another report from Argentina is even more remarkable. In 1989, Captain Ruben E. Lianza, a pilot in the Argentine Air Force, noticed a striking chain of aligned, very elongated craters in what is otherwise a nearly featureless farming district in the Pampas. The three largest craters in the line are all roughly 4 kilometers long and 1 kilometer wide. As at Campo del Cielo, there is a strong gradation from the largest craters at one end of the 30-kilometer chain to the smallest at the other end, but the rim profiles of the craters show an exact inversion of the trend at the other Argentine site: clearly the largest craters are the first in the series. The smaller craters were evidently made by fragments hurled downrange, in a spray about 2 kilometers wide, by the largest impacts. Impactite glasses and a small piece of a stony meteorite were found in the largest craters. The age remains poorly known, but the freshness of the craters implies an age of no more than a few thousand years. Reconstruction of the circumstances of the impact by Peter H. Schultz of Brown University attributes the crater chain to a near-horizontal impact by a stony asteroid with a diameter of 150 to 300 meters. The kinetic energy expended in excavating the first (largest) crater was about 350 megatons, and the total energy release of the impact was probably close to 1 gigaton (1,000 megatons) of TNT.

Thus near misses of comets and asteroids are well documented by modern observations, and grazing impacts also occur. The importance of Spacewatch and other asteroid search programs becomes more evident as we learn more about the impact hazard.

An interesting fringe benefit of Spacewatch's NEA detection operation is the discovery of other bodies of great scientific interest. The program has, not surprisingly, found comets. It has also found two Centaur family asteroids in orbits that cross the orbits of several of the Jovian planets. The prototype of the Centaur family, Chiron, is a giant ice-plus-rock cometary body with just the faintest trace of a coma and tail, with a solid nucleus a stunning two hundred kilometers in diameter. Computer orbital evolution studies of Chiron show that orbits of this class, which are frequently and violently disturbed by the gravitational effects of the giant planets, are generally chaotic. Over a time as short as one hundred thousand years, Chiron's orbit is knocked about so severely that it evolves into an Earth-crossing orbit. If the population of Chiron-type objects is about one thousand bodies, then one may move into an Earth-crossing orbit about once per thousand years.

At meetings on the impact hazard held in Tucson, Arizona, and Erice, Sicily, in 1993, extensions of the Spacewatch program into a global early-warning network were discussed by representatives of a dozen nations. Spacewatch itself is developing in two directions; to build a 1.8-meter (72″) second-generation telescope with four times the light-gathering power of the present instrument, and to assist in the construction of a companion observatory in India. But more telescopes are needed to provide continuous coverage of the night sky at all latitudes. A network of at least six or seven telescopes, all equipped with the latest CCD detectors and computers, would be required.

Such a network, called Spaceguard, was described in the 1973 science-fiction novel *Rendezvous with Rama,* by Arthur C. Clarke. In acknowledgment of Clarke's foresight, the astronomers of the world propose naming their global warning network Spaceguard. We can only hope that the funding for the real Spaceguard is not secured the same way that it was in Clarke's book. Clark has a small asteroid enter the atmosphere over eastern Europe on September 11, 2077. The brilliant daytime fireball begins to break up over Austria before impacting in northern Italy, destroying Padua and Verona and killing 600,000 people. The scenario is chillingly plausible. From the perspective of twenty years later, I would propose only two changes in the scenario. First, I would change the date to June 30 as a reminder of Tunguska and the Taurid complex. Second, I would have us build Spaceguard by 1997. That would give us an eighty-year head start on finding the hazardous bodies.

. . . .

Near-Earth Asteroid Population

Shoemaker (1983), updated by Lewis (1993)

Asteroid Class	No. Known (10/92)	Projected diam>1 km	Projected diam>0.1 km	Known	Number Easier to reach than Moon**	
					Projected diam.>1	Projected diam.>0.1
Aten ($a^* < 1.000$)	14	150	45,000	3	30	9,000
Apollo ($q^\dagger < 1.017$)	113	1,100	300,000	9	200	60,000
Amor ($q < 1.300$)	89	750	220,000	11	150	45,000
TOTAL	216	2,000	565,000	23	380	114,000

*a is the mean distance of the object from the Sun in astronomical units (Earth's distance)

†q is the perihelion distance (minimum distance from the Sun) in astronomical units

** An outbound velocity change of 6.0 km/s is sufficient to reach the Moon from low Earth orbit (600 km above the ground).

7
THE BASHFUL FACE OF MARS

Mars . . . experiences many localized dust storms, and many were observed by the two Viking orbiters. But the most spectacular meteorological events are those large storms during which the atmosphere over essentially the entire planet appears to fill with dust for periods of at least a few weeks or longer.

DR. CONWAY LEOVY, 1988

We suspect that the atmospheric pressure is just barely large enough to permit the development of global dust storms on Mars, as suggested by the confinement of their occurrence to times close to perihelion.

DR. JAMES B. POLLACK, ET AL., 1983

Earth's orbit around the Sun is quite close to a circle. At its greatest and least distances, it varies by less than 1.7% from its mean distance of 1.000 AU from the Sun. In contrast, the orbit of Mars is relatively eccentric, varying upward and downward from the mean distance of 1.524 AU by over 9.3%. As Mars ranges from its minimum (perihelion) distance of 1.382 AU out to 1.666 AU from the Sun (aphelion) it is climbing in the Sun's gravity well. Naturally, Mars slows down as it approaches aphelion. Likewise, as Mars falls from aphelion to perihelion it gains considerable extra speed.

Earth, being closer to the Sun than Mars, must travel faster to maintain its near-circular orbit in the stronger gravity field. Further, the distance around Earth's orbit is shorter than the distance around Mars's orbit. Thus Earth progresses quite a bit more rapidly around the Sun than Mars does.

About every 25.6 months, when Earth "laps" Mars on its annual trip around the Sun, the distance of Mars from Earth is at a minimum. That minimum can be as small as 0.365 AU, when Mars is at perihelion and Earth is at aphelion simultaneously at the same point on their orbits. At present, however, Mars's perihelion and Earth's aphelion do not line up in this way, and the planets do not get quite this close. The farthest the two planets can be from each other as

Earth passes between Mars and the Sun is 0.683 AU, when Mars is at aphelion.

Astronomers interested in observing Mars of course prefer to observe near the time of Earth's passage between Mars and the Sun, when the distance between the planets is small, and Mars is spectacularly large and bright to view. This time, when the Sun and Mars are opposite each other in the sky, is called opposition by astronomers. But, because of the eccentricity of Mars's orbit remarked above, not all oppositions are equal. At oppositions that occur when Mars is near perihelion the apparent diameter of the planet as seen from Earth is nearly twice as large as at the least favorable oppositions. The apparent area of the disc of the planet is then nearly four times as large at the best oppositions as it is when Mars is near aphelion at opposition time. Observers interested in mapping the surface or analyzing the atmosphere of Mars obviously prefer to observe near the time of opposition. The demand for telescope time is by far the most intense, however, at the times of the most favorable oppositions.

Spacecraft launched from Earth to Mars on a minimum-energy (maximum-payload) trajectory must be launched from Earth into an elliptical orbit around the Sun that grazes Earth's orbit at the point of launch, and grazes the orbit of Mars at the point of arrival. A spacecraft on a minimum-energy trajectory always arrives at Mars exactly halfway around the Sun from the point of launch. Therefore any spacecraft launched on an optimum orbit to Mars must be launched when Earth is at the point on its orbit exactly opposite to the point on Mars's orbit where Mars will be at the time of the arrival of the spacecraft. These ideal launch times are naturally spaced about 25.6 months apart in time. Launches more than a couple of weeks earlier or later than that optimum date must follow less-than-optimum orbits: a given rocket cannot carry as much payload if it is launched more than a few days earlier or later than the optimum date. The interval of time, usually less than a month, within which a rocket launch is capable of reaching a target planet with a non-zero payload is called a launch window.

But, because of the great variation in Mars's distance from the Sun, some launch windows may correspond to arrival at Mars when Mars is near perihelion; others, near aphelion. It costs much more energy to reach Mars when it is farthest from the Sun. Thus launch windows to Mars differ widely in their attractiveness. A Mars launch window opened in November of 1992, but with an unfavorable Mars arrival close to its aphelion distance. In January 1997 and February

1999 there are Mars launch windows for missions that reach Mars slightly before and slightly after two different perihelion passages, very close to the July 1986 near-optimum alignment. At these times, any given booster rocket can launch the heaviest possible payloads to Mars.

These moderately arcane constraints on the scheduling of astronomical observations and spacecraft missions to Mars have, as it happens, some very important ramifications. Not only are astromonical observations of Mars very strongly clustered around the date of opposition, but the most ambitious spacecraft missions to Mars (that is, those that carry the greatest weight of payload) must be launched to arrive at Mars close to the time of perihelion passage. The good news is that, for a few precious weeks, we have superb correlated astronomical and spacecraft data about Mars. The bad news is that, for most of the Mars year, Mars is so small and faint in Earth's sky that hardly anyone bothers to watch it—and those who do cannot expect to see much. Also, spacecraft with limited lifetimes will return virtually all of their data at and slightly after the time of perihelion passage. Most of the Mars year will evade study.

The first astronomer to study the surface of Mars was Galileo Galilei, who noted the phases of the planet in 1610. Working with a better telescope, the Dutch physicist Christian Huygens in 1659 recognized permanent dark markings on the Martian surface. The Italian astronomer Giovanni Cassini, working principally in France, used his observations of these markings over several oppositions to calculate a rotation period of twenty-four hours and thirty-seven minutes for Mars. Generations of observers slowly improved their sketch maps of Mars. The permanently dark areas, thought to be oceans, were called maria (seas), just as on the Moon.

Improvements in techniques for designing and building large telescopes slowly increased the size, power, and optical quality of the largest telecopes until, in the nineteenth century, many good-sized telescopes were available in many nations. These telescopes revealed ice caps at both poles of Mars and documented seasonal changes in color and contrast. Most consistent was the "wave of darkening" that spread out from each ice cap during local spring, a phenomenon strongly suggestive of seasonal changes in vegetation.

As the world was puzzling about these seasonal changes, an Italian Jesuit priest who was also an astronomer, Angelo Secchi, reported in 1869 that he had seen, at times of the most superb stability of the atmosphere, faint, dark lines on Mars connected with the large dark areas. Assuming that they were narrow features associated with water, he called these features by the Italian word for channels (*canali*). This word was widely translated into English as "canals," a word with clear implications of artificial origin. Several astronomers reported that, on rare occasions, they could see something akin to these canals. Many other astronomers with at least equal experience denied that there were any such features. The level of discourse, however, was soon to be irrevocably altered. In 1895 a Boston-Brahmin-turned-astronomer, Percival Lowell, published a popular book filled not only with his own observations of the elusive canals made from his personal observatory in Flagstaff, Arizona, but with his personal interpretation of their significance. In short, he concluded without reservation that the canals were artificial constructs of technologically advanced alien beings. The canals, Lowell announced, were designed for the purpose of distributing melt water from the polar caps to a rapidly desiccating world.

Lowell's book directly and powerfully stimulated the public imagination. That impact was greatly amplified by the powerful fictional treatment of Lowell's Mars by the English journalist and author Herbert George Wells. At the age of twenty-nine, Wells had just published his first novel, *The Time Machine*, when Lowell's book appeared. Wells, intrigued by the idea of a dying Mars peopled with an ancient intelligent race, adopted this idea as the central premise of his novel *The War of the Worlds*, first published in 1898.

Modern studies of Mars confirm the seasonal variations of the surface brightness, but without biological intervention. The culprit is dust, not vegetation. With rising spring temperatures, dust, which has accumulated upon and brightened dark basaltic rocks during local winter, is stirred up by the faster-moving winds. The dustiness of the atmosphere increases as the winds sweep the volcanic rocks clean. Stronger heating further fuels the atmospheric heat engine: on Mars, the very strongest heating occurs near the time of perihelion passage. Therefore, for sound physical reasons, Mars tends to have its most spectacular, often global, dust storms near perihelion. When Mars is easiest to observe from Earth, it often hides its surface beneath a thick veil of windblown dust, so opaque that only a few of the highest mountaintops can be discerned through it!

The two Viking Lander spacecraft, principally instrumented to carry out a search for evidence of life on Mars, both carried small meteorological instrument packages. Both landers also survived long enough to experience dust storms. Further, the two Viking Orbiter spacecraft returned detailed photographic coverage of the raising of dust by giant dust devils, some of them several kilometers high and hundreds of meters in diameter. Meanwhile, local changes in the surface were monitored by the Lander cameras. Comparison of the Orbiter and Lander data permits a detailed portrait of the origin and evolution of a giant dust storm.

First, as Mars coasts closer to the Sun, approaching the perihelion point on its orbit, and solar heating of its surface becomes more intense, rising currents of heated air (called thermals) become ever more important as carriers of heat from the surface. Under the influence of the Coriolis force, these rising currents begin to form spirals, which assists them in maintaining their cohesiveness. The rising spirals of hot air, as they gain more and more energy, generate higher and higher wind speeds at the surface. Soon, the wind begins to move small surface grains. The grains most easily mobilized are small but not as fine as dust. As these grains begin to move and bounce off the surface, a process called saltation, each bounce dislodges a blizzard of tiny dust particles which, once airborne, take a long time to fall to the ground. But the wind that they encounter is not simply a horizontal breeze; it is the outer surface of a hot, buoyant, rotating dust devil. The rising, rotating column of air rapidly whisks the dust up to altitudes of several kilometers. At times of strong solar heating there may be many active dust devils at the same time. The intensity of heating culminates at perihelion, when dust devils become very numerous. The dust raised reaches the stratosphere, but the rising column of hot air loses buoyancy when it reaches the lower stratosphere and ceases to rise. Deprived of its energy source, and left without the organizing effect of the Coriolis force, the dusty air from the core of the dust devil stops rising and spreads horizontally. In a great perihelion dust storm, the stratospheric dust clouds from countless dust devils overlap and shadow the surface, cutting down on the amount of sunlight reaching the ground by reflecting much of it away into space at altitudes far above the ground. The dust therefore cuts off the ultimate source of the disturbance that raises more dust. Wind speeds at the surface drop dramatically, no new dust is raised, and the surface temperature drifts downward. Cooling of the surface at night, which normally occurs through radiation of heat into space, is hampered by the dust screen overhead. The day-night temperature swings, usually very wide on

Mars, become much more moderate. Then, over a period of weeks, the dust slowly settles to the ground. The dust storm, which may have begun very abruptly, fades very slowly away.

The circumstances on Mars, of course, are very different from those on Earth. Mars has a very tenuous atmosphere that is so transparent that the surface cools off dramatically at night. Desert areas on Earth often have wide diurnal temperature ranges due to nighttime radiative cooling through very clear skies. At my home in Tucson, summer days that reach 110°F may be followed by nights that drop to 70°F. This is a cooling of 22 Celsius degrees. By contrast, the Martian atmosphere is so thin that summer equatorial temperatures near perihelion may drop from an afternoon high of nearly 20°C to a nighttime low of −80°C, a diurnal temperature range about four times as large as we ever see on Earth. (*Why* the atmosphere of Mars is so thin is a fascinating question that we shall return to in chapter 12.) Earth also has abundant water, both as vapor and clouds in the atmosphere and as oceans, lakes, and polar ice. Water contributes cloudiness and an infrared screen that help seal in surface heat at night, cloudiness to help exclude solar heating in the daytime, and massive transport of heat by atmospheric and oceanic circulation. Further, water in the form of rain and snow is very efficient at cleaning dust from the lower atmosphere. Clearly we cannot simply import our *observations* of Mars to Earth. Instead, we take the understanding of the laws of nature that we glean from our experience in the Martian environment and apply that theory to processes in the very different environment of Earth. It is the laws of nature, not the local manifestations of them, that are universal.

When this new insight into the behavior of dust was brought home from Mars, it was natural to wonder whether it could help us understand dust storms on Earth. Driven by the curiosity natural to people who spend their lives trying to figure out how things work, a serious attempt was made by a group of planetary scientists to use on Earth this insight into the behavior of planets gained from the study of Mars. The effort was successful, and, as so often happens, the new understanding illuminated an area of human activity so remote from Mars science that no one would ever have thought to link them in advance.

The dust theory imported from Mars showed that it was possible, under the most extreme of circumstances, to generate a global dust shroud on Earth. Not in a billion years of winds and storms would such an event ever arise from natural atmospheric processes. But if some event were to inject billions of tons of dust into the atmosphere quite suddenly, then global heating patterns would be disrupted, and the circulation of the atmosphere would be altered dramatically, planetwide.

The authors of that study, Owen Toon and a team of colleagues under the gloomy acronym TTAPS (for Toon, Turco, Ackerman, Pollack, and Sagan), found one dismayingly plausible way to inject the necessary mass of fine particles into the stratosphere: nuclear war. Surface bursts of large nuclear weapons are an essential part of strategic nuclear war. In view of our earlier discussion in chapter 4 concerning the optimum heights for explosions, this may seem a puzzling state of affairs. But in fact the very first priority of a nuclear aggressor is to pull the teeth of his adversary. This means targeting the enemy's strategic missiles and bombers with the first missiles launched. This is technically known as a counterforce targeting strategy. The enemy strategic assets will largely fall into three distinct categories. The first, submarine-launched ballistic missiles (SLBMs), are almost impossible to target because missile-firing submarines are very hard to find. The second type of target is strategic bomber bases. Airbursts are extremely effective against unhardened bomber stands. At least one surface blast is required per base, however, to crater the main runways so deeply that bombers protected by hard stands will be denied the ability to take off. These surface bursts raise vast amounts of dust. The third class of strategic targets is hardened ICBM bases. Missiles in deep, shock-isolated silos are extremely demanding point targets. Airbursts are ineffectual against silo-based missiles (except, of course, while they are in flight after launch). The only way to prevent launch is to target each silo and underground command post with large numbers of very precisely targeted and rather large warheads (one hundred kilotons or more). These warheads are fused to detonate below ground level, in order to drive as much as possible of the blast wave from the explosion into the ground near a silo. The idea, of course, is to shake the missile in its silo so severely that its tanks rupture, its structure fails, or its electronics are disabled. The large majority of the counterforce explosions are subsurface bursts targeted against silo farms. They raise horrendous amounts of dust, which is carried by the rising mushroom clouds into the stratosphere. The TTAPS scenario predicts that even a limited nuclear exchange involving a few percent of the world's nuclear arsenals could lift enough dust to cause a northern-hemisphere nuclear winter.

The use of nuclear weapons against urban targets is almost inevitable in a nuclear war. Large numbers of important military installations are located in or immediately adjacent to urban areas. Nuclear strikes against cities obviously have disastrous effects on local populations: it suffices to read of the horrors of Hiroshima and Nagasaki after twenty-kiloton nuclear airbursts and to try to imagine what multiple twenty-megaton explosions would do. But the realization of Toon and his collaborators was that urban areas are vast concentrations of flammable materials. Gasoline, aviation fuel, heating oil, and diesel fuel tank farms are always located near centers of demand. Cities are full of gas stations, motor vehicles, natural gas pipelines, and the like. An airburst ignites flammable materials out to distances of many kilometers from ground zero. The blast wave from the explosion ruptures storage tanks and gas pipelines and overturns and ignites vehicles. The black sky and sea of flames described by the survivors on the outskirts of Nagasaki will be multiplied by a hundred in area for each city struck by a large thermonuclear warhead, multiplied again by a hundred cities, and multiplied yet again by a factor of several to allow for the much larger concentration of flammable materials per square kilometer in a high-rise modern city. According to the model, the dust raised by a thousand megatons of explosions (the present inventory of nuclear weapons on Earth is some twenty thousand megatons) would easily suffice to blacken the skies of the entire northern hemisphere with dust, smoke, and soot. For a nuclear exchange that principally targets urban areas, total explosive yields as small as five hundred megatons could cause global climate changes. The Sun would be shut off for months, and temperatures would plummet, especially in continental interiors, far away from the moderating effect of the oceans. The result, in short, is a months-long nuclear winter.

The recognition of the nuclear winter threat, which arose out of research conducted for totally unrelated purposes, is a typical example of the kinds of benefits realized by basic research. In practical terms, the nuclear winter scenario has proved very useful in motivating politicians to think about and advocate nuclear nonproliferation policies and strategic arms reduction initiatives. American Mars science has made our daily lives a little safer; it has motivated the leaders of not just the United States, but also Russia, Belarus, Ukraine, and Kazakhstan, the heirs of the Soviet Union's nuclear arsenal, to reduce dramatically, and even eliminate, their strategic nuclear weapons.

But we know another way, besides nuclear Armageddon, to provide a thousand megatons (one gigaton) of explosive power at ground level: comet and asteroid impacts. Volcanic explosions larger than about one gigaton may be impossible: Krakatau weighed in somewhere around one hundred megatons, and the explosion of Thera on the Mediterranean island of Santorini (the probable source of the myth of the destruction of Atlantis) may have been close to one gigaton. One-gigaton impact explosions occur with a frequency of one per ten thousand years. Dust raised by explosions of this size on a continental surface is a serious threat to life over at least a hemisphere, and possibly over the entire world. But over 70 percent of the impacts on Earth occur in deep water. These of course raise little or no dust unless the impactor reaches the ocean floor before detonating. The question of what happens when a large impactor strikes water is an interesting one, and we will return to this matter in chapter 12. But now we know that giant explosions on land can indeed have profound effects on climate through injection of dust into the stratosphere. What use we can make of this knowledge remains to be seen. We also know that impacts very much larger than a gigaton are possible, with devastating consequences. Therein lies the tale told in chapter 8.

8

ENDS OF GEOLOGICAL AGES

Geology was liberated as a science by Hutton and Lyell at the beginning of the last century by means of the great principle of "uniformity." At the time that it was enunciated, it was the most important single principle to emerge in the history of our science and, in a strictly limited sense, is equally true today. Following the overthrow of catastrophism, however, there has been a natural tendency to over-compensate and to avoid catastrophic interpretations even when the evidence called for one. Our science is not to be held back by rigid application of an all-encompassing principle under every circumstance. The increasing demonstration of violent events in our past environment is supported by the discovery of craters on the other side of the Moon and on the surface of Mars. That is, in all parts of the solar system accessible to close inspection. . . .

I shall, therefore, land a large or very large meteorite in the Paleozoic Pacific at the close of the Frasnian. . . .

Dietz suggests that a giant meteorite falling in the middle of the Atlantic Ocean today would generate a wave twenty thousand feet high.

This will do.

PALEONTOLOGIST DEWEY J. McLAREN, 1970

The geological history of Earth is read from sediments. Each page of the book is a layer of sediment inscribed with a wealth of information written in the languages of chemistry, physics, and biology. Most sediments from ancient times have been processed by natural geological processes, heated and compressed by deep burial within the crust, weathered away by the erosive action of wind and wave, chemically attacked by water and oxygen, or even melted completely into a magma within which the historical information has been destroyed. Fortunately, some sediments, by rare good fortune, survive from very ancient times with little or no loss of legibility. But in any locale, only pieces of the textbook remain intact. Any particular location may have collected sediment for part of the time, but suffered erosion the rest of the time. An erosive environment not only prohibits new sediment from accumulating, but also destroys the topmost (latest) pages from the history book already laid down. Each local sample of

sediment therefore is chronologically incomplete. In one place we may find a stack of pages from page 22 of chapter 1 through the second page of chapter 3, with a few pages from the middle of chapter 12 on top of them, surmounted by the end of chapter 38 and the complete texts of chapters 39 through 46, right up to the present. Local processes of course are different at different places. Erosion and deposition begin and end at different times at sites hundreds of kilometers apart. Therefore the texts recovered from deep columns of sediment in different locales sample different parts of the book. It is the task of the geologist who wants to read the book of Earth's history to correlate the records preserved at many widely separated sites and interleave these records to produce a complete history. This science is called stratigraphy, literally, "the writing of the layers."

As the stratigraphers of the eighteenth century built up their book of history, they found many clues to assist them in their immense task. First, there were the fossils. Over a lengthy period of recent history, sediments commonly contain the remains of ancient plants and animals, the proper domain of paleontologists. As on the modern Earth, some species are found only locally, some over an entire ocean basin, and yet others may be found very widely over the entire planet. The latter, which evidently find themselves at home almost everywhere, are appropriately called cosmopolitan species. In shallow marine sediments from anywhere on the present Earth one might expect to find the skeletal remains and teeth of sharks. Certain mollusks might be found only in Florida; others only in the Red Sea; still others only in Hudson Bay; or some might be found in all the tropical and subtropical oceans of the world. Of course, remains of large creatures are much less common than the remains of small ones. One might comb thousands of square kilometers of shallow-water offshore sediments before finding a single shark skeleton, but a single square meter may yield hundreds of mollusk shells. This principle extends down to even smaller sizes: by far the most abundant fossils in the world are tiny, even microscopic.

It is in the world of micropaleontology that stratigraphers find most of their information about the succession of ancient life forms. What they find on their Lilliputian scale, with vast populations of fossils at their disposal, is also reflected faithfully (albeit with greatly inferior statistics) on the normal level of size. The history of life, however we read the book in the sediments, has been severely disrupted many times. The uniformitarian's slow, plodding, stepwise

progress through an endless succession of minute variations, with vast expanses of time spent in leisurely competitive winnowing, does indeed go on most of the time. But, contrary to the Darwinian concept of gradual origin and decline of species, almost all evolutionary change occurs during a few brief episodes that occupy only a tiny fraction of the time. Gradual ecological changes, resulting from single-point mutations, climate change, and competition, are punctuated by a number of violent and very brief extinction events. Vast numbers of species and genera disappear abruptly in concert, despite having nothing in common in terms of proximity of habitats or environmental preferences (favorite temperatures and salinity; taste in food; sunlight requirements, etc.). These global extinction events occur at well-defined discontinuities in the stratigraphic record: the biological and geological histories of the most widely separated parts of our planet record the same major events.

The greatest recorded extinction event occurred 225 million years ago at the end of the Paleozoic era ("ancient life"), with the closing of the Permian period. This was the time of transition from simple early reptiles such as *Seymouria, Eryops, Dimetrodon,* and *Edaphosaurus* to the true dinosaurs. (Readers unfamiliar with these creatures should ask their children for further information.) For the next 160 million years, throughout the Mesozoic era ("intermediate life"), the dinosaurs ruled Earth. In the early Mesozoic, during the 35 million years of the Triassic period, the dinosaurs proliferated, diversified, and established their mastery. The end of the Triassic was marked by what is probably the third largest extinction event since the Cambrian. Next, during the 54 million years of the Jurassic period, many of the most familiar species of large dinosaurs, including giant sauropods like *Diplodocus* and carnosaurs such as *Allosaurus,* prospered, laying the groundwork for ambitious movie ventures of the distant future.

Then followed the long afternoon of the dinosaurs, the 71 million years of the Cretaceous period. Throughout the last third of the Cretaceous the dinosaurs slowly declined in diversity and numbers. The end of the Cretaceous was marked by the second-largest mass extinction of all time. Over 90 percent of the species then living on Earth vanished abruptly and nearly simultaneously. The final fall of the dinosaurs at this time was only a tiny part of the extinction story, but has endless appeal in the minds of children and former children alike. Their great size lends them an aura of invincibility that we do not instinctively attribute to much more robust species that happen

to be microscopic in size. Perhaps the most frequently asked question among students of geology is, "What killed the dinosaurs?" There are surely many answers to this question, not one of which is impeccably established by hard evidence. We will instead ask the much more interesting (and more answerable) question, "What happened to life on Earth at the end of the Cretaceous?" The answer to this question is of personal interest to all intelligent mammals, since the ascendancy of mammals dates from that very collapse of the Cretaceous fauna. The end of the Mesozoic era and the beginning of the Cenozoic era ("recent life") coincide with the end of the Cretaceous. In the wake of the extinction, a vast range of ecological niches were opened up abruptly in the early Tertiary period (the earliest Cenozoic) to innovative species that could quickly adapt to the new world ecology. Our remote ancestors were among those who found it expedient to change and diversify.

For those inclined to feel smug about the great imperial success of the mammals in general, and of human beings in particular, it is enlightening to reflect on the time scale of our species. Our first recognizable human ancestors date back only about 2 million years, and mankind cannot fairly be said to have risen to domination of the planet more than a few thousand years ago. Perhaps we should frame the rise of mankind in terms of the rise of agriculture, writing, and cities some six thousand years ago. In humbling contrast, the dinosaurs ruled the planet for 160 million years, some thirty thousand times as long as human culture has so far existed.

When the long-delayed demise of the dinosaurs finally occurred, the crash was spectacular. Of all the global discontinuities in the geological record, the best-studied has been the end of the Cretaceous (K) period and the beginning of the Tertiary (T) period. (The letter C had already been appropriated by the Cambrian period.) Although the Cretaceous extinctions were not the largest in biological history, the event was so recent (only 65 million years ago!) that evidence is relatively easy to gather. The Cretaceous/Tertiary (K/T) boundary layer tells us an astonishing story.

When geologists read the pages just before the Cretaceous-Tertiary boundary they find the score of a typical uniformitarian opera; a long list of overly familiar characters playing tiny variations on a thoroughly familar story, with arias and recitatives that all somehow inspire déjà vu, running on and on with soporific effect. But suddenly, marking the transition to the Tertiary, there is a sharp global discontinuity in everything. The sediments change from the local norm, found in hundreds of variations throughout the world, reflecting the humdrum processes of the last few million uneventful years. Suddenly, over the entire world, at every site where ocean or lake sediments have been preserved, there is a millimeters-thick layer of gray clay. The thickness of the layer varies little from place to place, usually between one and ten millimeters, and its composition is strikingly uniform, largely unrelated to the local chemistry. Only in the vicinity of Mexico and the Caribbean Sea is the layer systematically thicker. Worldwide, this layer contains a staggering several trillion tons (10^{18} grams) of clay, or about 1,000 cubic kilometers of material. The thinness of the layer, and its uniformity, suggest that it was deposited in a single event, with a duration possibly less than a year. In this layer, physicist Luis Alvarez and his colleagues in 1980 found an enormous concentration of certain metals that are very rare in Earth's crust, notably the platinum-group metals. The first of these metals to be documented in the K/T boundary layer was iridium, but we now know that a number of other metals are also enriched. Interestingly, the pattern of relative abundances of these metals is very different from that found in Earth's crust, but indistinguishable from the pattern of average solar system abundances. The pattern is not similar to any terrestrial source, but matches average meteorite material very well. It is, in effect, an alien fingerprint. Mixed in with the global iridium-bearing clay layer are tiny particles of heavily shocked quartz. In addition, in the Caribbean basin, the boundary layer is thicker and contains larger glassy spherules and droplets that have been largely, but not completely, altered by seawater.

Above the boundary, the world quickly quiets down to a new mode of functioning. As the transients die out, a sadly reduced cast of characters improvises its way into a new stable ecology. The intensity of the experience dwindles again as the inquiring geologist, like the audience at the premiere of *Madama Butterfly*, begins to notice apparent quotations and paraphrases. The tendency to stand as they did and shout, "Give us something new!" bubbles up inside us. Uniformity is back.

It is as if a score by Philip Glass were inadvertently printed by a sloppy publisher with a page of the *1812 Overture* inserted. The orchestra, discovering the error after the first cannon blast, struggles to regain control of the concert. Meanwhile, the cognitive dissonance of the experience should shock any uniformitarian in the audience fully awake.

Conservative elements steeped in uniformitarian philosophy cannot be expected to welcome such a rude awakening. The critics rose in rebellion. Devastatingly sarcastic anticatastropic polemics, reminiscent of some of the early reviews of Tchaikovsky, were even published as editorials in *The New York Times* on February 17, 1981, and April 2, 1985. Both diatribes make hilarious (and embarrassing) reading from the perspective of the 1990s. The richest touch was the dictum in the latter that "Astronomers should leave to astrologers the task of seeking the cause of earthly events in the stars."

If the resurrection of this slogan does not clearly expose the folly of its anonymous authors and bring a rosy flush to their pallid cheeks, *nothing* will. We are transported right back to 1660, with the witch-hunters burning harmless, eccentric old women and the theologians debating whether comets and fireballs were celestial portents. But in 1985 the Inquisitors were paleontologists. Once cast in the role of Guardian of Truth and Traditional Wisdom, a scientist ceases to be scientific. He becomes yet another defender of dead scholasticism, a mere hawker of dogma.

The paleontological perspective on the K/T boundary is of singular interest. Here an entire profession of experts trained in uniformitarian principles stares catastrophe in the face. Some, of course, react with denial. Their responses go something like this: "catastrophes are impossible; it must have been a terrestrial event, like a volcanic eruption; well, a *lot* of volcanic eruptions really close together in time; sort of like a terrestrial catastrophe, but not really; and the iridium layer must be from volcanoes, but it was subjected to unknown chemical processes that happened to make the metals look like meteorite material; of course volcanoes can be violent, so it is reasonable to suppose that the glass and shocked quartz are made by unknown volcanic processes. It's true that impacts like Meteor Crater do really occur, but they are too small to have any global effect; there is no evidence for much bigger impacts because the big craters are cryptovolcanic; and anyhow those giant impacts had no effect on life; so the mass extinctions were due to terrestrial causes; of course, there weren't really any mass extinctions; it just looks that way; besides, we know that the dinosaurs were already extinct by that time, and there were clearly a lot of dinosaurs still around a couple of million years later." This school of thought, although expressed with splendid passion and eloquent conviction, somehow seems to lack the power to convince the audience.

Other paleontologists are far more ready to look the K/T boundary in the face. They know a catastrophe when they see one. To them, the question is one of mechanism: exactly what happened, and what were the killing mechanisms? This issue was first raised in a much more limited context by two geologists, Allan O. Kelly and Frank Dachille, in their 1953 book, *Target: Earth*. Their book, taking as its foundation the then-widespread belief that continents did not drift, addressed the evidence for gross movement of Earth's magnetic poles every million years or so by proposing that comet and asteroid impacts frequently "knock the Earth off its axis." A simple calculation would have revealed that any impact with enough energy to change Earth's axis detectably would have completely overwhelmed the biosphere, probably sterilizing the surface of the planet. Also, we now know that the cratering rate on the planets must be at least ten thousand times lower than Kelly and Dachille supposed. However, their scenario still has its points of interest: they suggested that these abrupt axial shifts literally displaced the oceans from their basins and caused dramatic and virtually instantaneous climate changes near the poles. In keeping with their belief that such events are very common, they traced back both the wide diversity of myths about floods and deluges from many cultures throughout the world (Judeo-Christian, Babylonian, Hindu, Chinese, Chaldean, Greek, Amerind, etc.) to the most recent such event. They then linked the extinction of the woolly mammoth, woolly rhinoceros, saber-toothed cat, giant ground sloth, and others to an impact eight thousand to fifteen thousand years ago. But they did not connect their ideas to the much earlier extinctions recorded in sedimentary rocks.

Interestingly, the first scientist in modern times to propose that the great mass extinctions in the geological record were caused by an impact was a paleontologist, M. W. de Laubenfels. In 1956 he appealed to the then-meager astronomical evidence for close flybys of Earth by asteroids, citing the 1937 approach of the near-Earth asteroid Hermes to within eight hundred thousand kilometers of Earth and studies of the cratering history of the Moon by American Nobel laureate Harold C. Urey. Writing in the *Journal of Paleontology*, de Laubenfels also appealed to the accumulating evidence on the devastation wrought in Siberia by the Tunguska explosion, bemoaning its inability to influence American paleontologists: "If this area had been in the United States, American scientists would have been impressed. They would not regard danger from impact as being preposterous, an assumption which is now common."

The same concerns arose in the mind of the eminent and brilliant

Estonian astrophysicist Ernst J. Öpik. A man of astoundingly eclectic interests, he contributed to the theory of stars, planetary evolution, impact phenomena, and many other areas while writing scientific materials fluently in Estonian, Russian, French, German, and English, and even publishing poetry in Irish. In 1958, in the course of studying both the dynamics of meteor flight in the atmosphere and the collision probabilities of nearby asteroids with Earth, he published an extended abstract on impact-induced catastrophes in the *Irish Astronomical Journal.* In addition to the lethal effects of large impacts, he suggested that sufficiently large impactors could penetrate the continental crust, triggering the formation of huge areas of flood basalts such as those in the Deccan Traps in India and the Columbia River basin in the American northwest. Thus massive volcanism might make more sense as a consequence of, not an alternative to, impact-driven extinctions.

Another paleontologist, Dewey J. McLaren, speaking from the top of his profession, proposed in his presidential address before the Paleontological Society in 1970 that global marine extinctions could be brought about by two known causes, dilution of the oceans by vast amounts of fresh water, or turbidity sufficiently intense to intercept sunlight. His remarks were aimed principally at the Frasnian extinction event in the Devonian period, when life was confined to the oceans, but they are equally relevant to the Cretaceous marine extinctions. McLaren concludes that the former mechanism is numerically untenable because there is not remotely enough fresh water and ice in the world. The latter seems to require a huge impact event. His remarks at the beginning of this chapter prove the openness of an outstanding paleontologist to new ideas. Many of his paleontology colleagues, however, were predictably outraged, and some remain so to this day.

Harold Urey, writing in 1973 at the age of eighty, proposed a specific relationship between heating of the biosphere by comet impacts and global extinction events. He also offered a tentative correlation between major formation events of glassy impactite material called tektites and the ends of several geological ages. Urey was aware of only one earlier appearance of this idea in print—an interview he himself gave in the *Saturday Review of Literature.* Urey, with tongue firmly in cheek, admitted that the interview might not have reached his scientific colleagues, since he was not aware of any scientist besides himself who read that journal. Nonetheless, the point deserves making: all of these pioneers, Kelly and Dachille, de Laubenfels,

Öpik, McLaren, and Urey, were writing in ignorance of each other's suggestions. Not one of these articles cites any of the others; indeed, de Laubenfels, writing the earliest on impact-caused ancient extinctions, does not cite any literature whatsoever in his paper. The reason is not hard to find: astronomy departments do not subscribe to the *Journal of Paleontology*, and paleontologists do not make up a large part of the subscription list of the *Irish Astronomical Journal*, which in any event is a very select fraternity. Urey, however, chose to publish in *Nature*, a very widely circulated British journal famous for quick publication of original ideas in all scientific disciplines (with all the implications, both positive and negative, of that distinction). As a direct result, Urey's paper is widely cited by latter-day catastrophists, whereas the others are cited, if at all, by narrower subsets of the field.

A number of paleontological studies of the Cretaceous extinction event have been stimulated by Luis Alvarez's discovery of the iridium-rich clay layer. The first lesson to be learned from these studies is that the extinctions were indeed devastating. According to a study by paleontologist J. J. Sepkoski Jr., there was a 35 percent loss of genera of foraminifera, bryozoans, corals, marine gastropods (snails), and marine vertebrates. (Of course, for a genus to become extinct, 100 percent of the species in that genus must be lost). Marine arthropods and echinoderms (sea urchins, sand dollars, etc.) lost only about 25 percent of their genera. Brachiopods and marine bivalves (clams) suffered a 55 percent loss of genera. The ammonoids (similar to the nautilus), which had been a dominant organism in the Cretaceous seas, were completely eradicated.

In response to the vast mortality, new species arose as the survivors adapted to new environmental conditions and a new roster of food sources, predators, and competitors. Mutations that would have conferred no useful advantage only years earlier suddenly took on new significance, allowing the branching of each surviving species into several differently specialized and noncompetitive new species. This behavior is decidedly non-Darwinian. Darwinian processes of slow innovation and slow fading of older, less competitive species to extinction thrive in a uniformitarian world; new paradigms are needed to cope with the world of rare, violent catastrophes. Such a paradigm is the idea of punctuated equilibrium, which owes its prominence in part to the eloquent advocacy of Stephen Jay Gould, and in part (as sometimes happens) to the fact that it makes sense.

The "equilibrium" in this phrase, which would promptly be renamed "steady state" by any thermodynamicist, is the Darwinian con-

. . . .

tinuum of slow evolutionary change. The distinction is not an academic triviality: a system is in equilibrium if its state variables *never change*, and if there are no gradients in any of the fields or variables and no fluxes (flows of matter or energy) driven by these gradients. In common language, an equilibrium system is *dead*. In a steady state, some state variables are maintained constant by the gradients in certain fields and the resulting fluxes of matter and energy. Steady-state systems are often far from equilibrium, have an energy flow through them, and often can tap great reserves of available energy. In a word, a steady state system is *lively*, even truly alive, like a Darwinian universe. The "punctuation" in Gould's phrase is of course provided by catastrophic disruptions that destroy all the delicate balances of fluxes and gradients that define a steady state.

The agent by which the mass mortality at the K/T boundary was inflicted remains unclear in detail, but there is no shortage of suspects. Alvarez and his coworkers pointed out that the immediate effects of the explosion, including the blast wave from the impact, the ignition of fires by the brilliant, hot fireball, and the scouring of the surrounding countryside by boulders ejected from the impact crater, tend to visit a small portion of the surface of Earth with overkill. The observed global extinctions require global effects of the blast. The most obvious culprit was the fine-grained boundary clay layer itself, which demonstrably was distributed globally. The layer contains enough dust to shut off sunlight from the surface of the planet for several months to a year. In the absence of sunlight, solar heating of the surface comes to a halt. Infrared cooling of the surface is also inhibited, but heat radiation is less affected by dust than visible light. Therefore a planetwide dust storm should lead to severe cooling of the continents. James B. Pollack of NASA Ames Research Center and his coworkers Owen Toon, Tom Ackerman, Chris McKay, and Richard Turco saw that the sudden onset and global extent of the dust shroud could be studied using tools already developed in the planetary program: "A good, physically appropriate analogy to such rapid spreading is provided by Martian dust storms, which grow from local to global proportions in only 1 to 2 weeks."

In their computer model of the K/T dust cloud, Pollack and his colleagues found that the dust shield would indeed suffice to darken the ground to the point of cutting off photosynthesis completely for many months. Cooling of the surface would persist for several months to a year, but the oceans, because of their enormous thermal inertia, would cool only a few degrees. The continents, however, might cool

. . . .

by as much as forty Kelvin degrees (seventy-two Fahrenheit degrees) below normal. This amount of cooling would put most of the land area of the planet below freezing and in darkness so complete that animals could not see. The first and most obvious cause of global extinctions is the dust layer itself: freezing temperatures and darkness, by stopping photosynthesis, cut off the food chain at its source. Hypothermia and starvation kill the creatures that live higher on the food chain.

Another direct and unavoidable consequence of large impacts was pointed out by myself, Hampton Watkins, Hyman Hartman, and Ron Prinn at a conference held just a few months after the Alvarez team announced the discovery of the iridium-bearing clay layer. The basic point is as simple as the dust scenario: high temperatures in the atmosphere, such as lightning discharges and nuclear weapons explosions, cause a small portion of the nitrogen in the air to "burn" to make NO gas. This gas is then rapidly oxidized to NO_2 and N_2O_4, which react with water or water vapor to make vast quantities of a mixture of nitrous acid (HNO_2) and nitric acid (HNO_3). We pointed out several features of the pattern of extinctions at the K/T boundary that suggested massive acidification of the oceans as a possible cause of the selectivity of the deaths. We also presented a crude calculation of the amount of nitric acid that might be produced by an impactor of such large size. More refined calculations by Kevin Zahnle have reduced the estimated amount of nitric acid produced, but this reduction has been partially offset by new and larger estimates of the size of the K/T boundary explosion and by the recognition of two massive new sources of acid rain from that event.

This "chemical warfare" scenario makes a number of interesting points. First, the nitrogen oxides are toxic to animals. Any creatures that breathe the surface air directly are subject to severe lung damage, lung edema, and death. These same nitrogen oxides, in high but apparently plausible concentrations, will defoliate plants. The higher oxides, produced secondarily, are powerful absorbers of visible light: a mixture of NO_2 and N_2O_4 is a deep reddish brown. This gas absorbs visible light so well that plants could not photosynthesize even if they somehow retained their leaves. The nitrous acid produced later in the reaction chain is toxic, corrosive, mutagenic, teratogenic, and carcinogenic. You wouldn't even like the taste. The nitric acid acidifies rainwater over the entire Earth to a pH of 0 to 1, as bad as the worst industrially generated acid rain pollution ever monitored anywhere. The surface layer of the ocean down to a depth of tens of meters becomes so acidic that carbonate shells dissolve and their occupants die. Carbonates destroyed by the acid rain release vast quantities of carbon dioxide into the atmosphere.

The "splash" of shock-heated atmosphere from the initial impact and the prolonged rising column of superheated air from the magma-filled impact crater would reach to high altitudes far around the world, dumping nitrogen oxides into the stratosphere. The nitrogen oxides would wipe out the ozone layer almost instantly. Interestingly, marine geochemist J. D. Macdougall has found that a vast quantity of strontium weathered from the continents was dumped into the oceans at the end of the Cretaceous. The most plausible explanation of this observation is an abrupt, massive, global acidification of rainwater.

Acid rain falling on the continents leaches toxic metals from the surface rocks and soil, according to follow-up research done by Ron Prinn and Bruce Fegley in 1987. They calculate severe leaching of such toxic metals as beryllium, aluminum, mercury, lead, and cadmium. These toxic materials are washed down rivers into lakes and oceans, contributing to freshwater and marine extinctions. This corrosive, toxic acid indeed turns the rivers to wormwood. Other less toxic or benign metals such as calcium, magnesium, and strontium are also efficiently dissolved.

A major cometary impact may eject and accelerate to high speed a mass of material a hundred times as large as the mass of the impactor. Some of the gases from the explosion and fireball may reach escape velocity. A much larger mass of material is hurled out at speeds sufficient to rise out of the atmosphere and travel long distances before landing. Ann Vickery and Jay Melosh of the Lunar and Planetary Laboratory of the University of Arizona have shown that the debris from a large impact will travel over great distances, reentering the atmosphere over the entire surface of the planet. The time needed for a piece of crater ejecta to travel to the opposite point on Earth's surface is forty-five to sixty minutes. Thus over the hour immediately following a major impact, high-speed reentering debris showers down from the skies at orbital speeds all over Earth and burns up. The energy released by the reentry of this debris can be estimated, and is large enough to scorch the ground anywhere (perhaps even everywhere) on Earth with heat loads sufficient to ignite forest fires. Indeed, one component of the global boundary clay layer is soot. The total amount of unburned carbon in the soot is about what would be expected if 50 percent to 100 percent of the forests in the world were to burn simultaneously. Thus the immediate result of a K/T size impact anywhere on Earth would be wildfire ignition over the entire planet. These firestorms would burn unhindered to the point of exhaustion of the fuel. Unavoidable side effects of continentwide wildfires include the darkening of the skies by billions of tons of soot, the release of a wide range of toxic pyrochemicals produced by partial combustion of vegetation, and the production of equally enormous additional quantities of nitrogen oxides made by the high temperatures of the flames. Only exceptionally cloudy, boggy areas might survive the intense heat radiation from the reentering debris. In the event of an impact in deep ocean, rock ejection would be minimized and water would be thrown great distances. Firestorms, if ignited, might be promptly drowned under a deluge of water. Much depends on the exact size, energy, and location of the impactor. It would be most useful to find the crater, since its size tells us the energy content of the explosion.

Such a gigantic event as the K/T bolide impact requires a gigantic crater, at least 120 kilometers in diameter. Until quite recently, no crater of even approximately the right size and age was known. One poorly dated impact scar, the 35-kilometer Manson structure in Iowa, seemed to be a candidate until late 1993, when a careful potassium-40 decay study revealed that it is 73.8 million years old, in clear conflict with the 65-million-year age of the K/T boundary. Many authors had noted that 72 percent of the impactors falling randomly on Earth would land in the ocean. But, because of continental drift and recycling of the ocean floor, most of the ocean floor of 65 million years ago no longer exists. Further, searches for very large craters in deep ocean water could be frustrated by sedimentary and volcanic refilling of the impact crater. Overall, there is about a 50 percent chance that the K/T crater should no longer exist.

By the late 1980s it had become clear that the abundance of shocked quartz grains in the boundary clay layer was higher in the Americas than elsewhere, suggesting closer proximity to their source. Since quartz is an abundant continental mineral, but present on the ocean floor only where it has been weathered off nearby continents, an impact site on land seemed probable. Fieldwork in Haiti by Alan Hildebrand, then a graduate student in the planetary sciences department of the University of Arizona, showed in 1990 that a K/T boundary deposit there, previously attributed to volcanic action, actually consisted mainly of altered tektites, and contained abundant shocked quartz, both indicative of impact rather than volcanic origin.

Several candidate impact sites in the Caribbean basin and Gulf of Mexico were then suggested and explored. Hildebrand, with colleagues David Kring and Bill Boynton, suggested in 1990 that the impact might be a large, roughly circular, sediment-filled basin on the north slope of the Yucatán Peninsula in Mexico. This basin, called the Chicxulub crater, formed on the continental shelf in shallow water. It is at least 180 kilometers in diameter; indeed, one researcher claims an original diameter as large as 300 kilometers. This size requires an impactor of at least 1,000 cubic kilometers, which excavated a crater with a volume of about 100,000 cubic kilometers. Much of the excavated mass was vaporized.

The surface rock struck by the Chicxulub impactor is itself of sedimentary origin. It contains abundant carbonates, and is dominated by the mineral anhydrite, which is simply calcium sulfate. Vaporization of tens of thousands of cubic kilometers of target rock releases vast amounts of carbon dioxide and sulfur dioxide into the atmosphere. Both of these substances have important effects on the atmosphere. Sulfur dioxide is a potent source of acid rain, probably even more important than the nitric acid produced by the impact blast wave. It also forms a very reflective sulfate aerosol haze in the stratosphere, which efficiently reflects incident sunlight away from Earth. Carbon dioxide is a greenhouse gas, which can cause long-term temperature increases after the atmosphere again becomes transparent, and for at least several centuries thereafter.

There is certainly no shortage of devastating phenomena associated with large impacts.

That impact events may play an extremely important role in the modern history of life on Earth seems clear; but the impact rate was much higher in the distant past. To what extent has the origin, stability, and fate of life on Earth been regulated by impacts? Impacts upon the early atmosphere, before the origin of life on Earth, must have generated simple organic molecules such as formaldehyde and hydrogen cyanide, which react with each other and with water to make a wide range of interesting molecules that can serve as the building blocks of life. Thus early impacts, like other high-energy processes such as ultraviolet light, cosmic ray irradiation, and lightning discharges, support the production of the essential components of life.

But once the first primitive, delicate life forms have arisen, these energetic processes become a hazard to them. It is trivially obvious that a virus will not fare well when struck by lightning; it is not the fate of single primitive organisms that concerns us. The problem is that a large enough impact can deliver enough energy to heat the atmosphere to sterilizing temperatures, raise the temperature of the oceans to the boiling point, and boil the oceans away completely. Even life huddled about hydrothermal vents on the abyssal ocean floor, perfectly sheltered even from events as cataclysmic as a nearby supernova explosion, would be destroyed. It seems most likely, in fact, that primitive life arose and was destroyed several times over by very large impacts.

As the planets grew toward their present size they swept the inner solar system free of orbiting debris, the impactor flux declined, and single impacts sufficiently large to sterilize Earth became ever more improbable. They are not *impossible* even today, but no such event has happened for over 4 billion years. The energy required to sterilize Earth is equivalent to a two-hundred-kilometer-diameter comet nucleus colliding at typical cometary speeds of about fifty kilometers per second. The largest known cometary nuclei presently in Earth-crossing orbits are about ten kilometers in diameter. Thus large impactors in the near future may wreak havoc on Earth, destroying many species and sterilizing vast areas with searing heat, acid rain, darkness, continentwide deep freezes, et cetera, but the probability of an impact utterly destroying life on Earth between now and the time the Sun dies is too small to be of immediate concern. We have better things to worry about. For instance, there is still the possibility of nuclear, chemical, or biological warfare, there are still diseases that defy control, and there are still those much smaller impactors that may wipe out only a few thousand species, including our own.

Even more probable than the extinction of humanity is a catastrophe that destroys our culture while leaving some humans still alive. The Greenland and Antarctic ice cores show that the last 2 million years of alternating ice ages and warm interglacial periods were actually much more complex: in the generally warm interglacial periods, violent, brief temperature drops of 20 degrees Celsius (about 36 Fahrenheit degrees) were common. Likewise, during the frigid eras of ice sheet advances, numerous brief episodes of extreme warming occurred. The climate in the 2 million years since the time of the earliest hominids has generally been wildly unstable, *except during the*

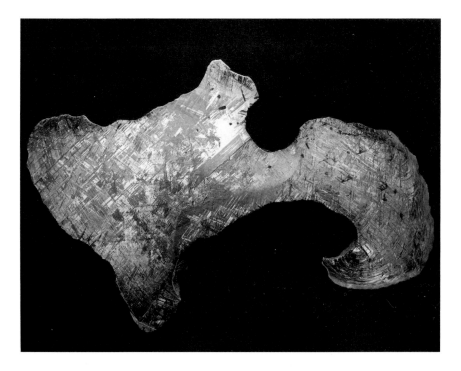

One of the Gibeon iron meteorites from Namibia. This piece of meteoritic iron has been cut, polished, and etched with acid to bring out its distinctive crystal structure. The far right edge has been severely deformed by a violent collision which has made the tough natural steel flow and bend.

Photograph courtesy of the Smithsonian Museum of Natural History

Meteor crater near Winslow, Arizona. This impact crater was formed by a tiny iron asteroid about 35 meters in diameter, or about the size of a ten-story office building. The crater is roughly 0.7 miles in diameter.

Photograph courtesy of NASA

The Leonid meteor shower of November 17, 1966, as photographed from atop Kitt Peak in Arizona. The original negative shows 70 meteor trails. The star images are trailed by the time exposure, and the radial streaks are the meteors. The two bright dots near the center of the photograph are entering head-on to the observer. At the time that this photograph was taken, visual observers were counting about 40 meteors per second.

Photograph courtesy of Dennis Milon.

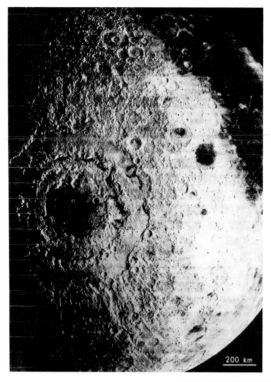

The Orientale Basin on the Moon as photographed by the Lunar Orbiter 4 spacecraft. The central basin, flooded with dark basaltic lava after an asteroid or comet impact, is about the size of Arizona or New Mexico. Texas would fit within the outer rings of mountains.

Photograph courtesy of NASA

The Caloris Basin on Mercury, another gigantic impact feature caused by a comet or asteroid. The point on Mercury's surface opposite this impact site was severely affected by the shock wave transmitted through the planet's core.

Photograph courtesy of NASA

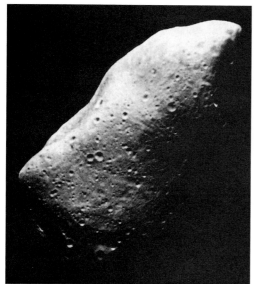

(LEFT) The Belt asteroid 951 Gaspra as seen by the *Galileo* spacecraft on its way to Jupiter. A heavy blanket of impact-crushed regolith covers its surface.
Photograph courtesy of NASA

(BELOW) The summit of the gigantic Olympus Mons volcano thrusts above the dusty, cloudy atmosphere of Mars. At times of great planet-wide dust storms, only a few of the tallest volcanoes can be seen poking through the global dust veil. Olympus Mons is about 26 kilometers (86,000 feet) high.
Photograph courtesy of NASA

Three large impact craters on Venus are visible in this radar image made by the *Magellan* spacecraft from orbit around Venus. Small craters are absent on Venus because small bodies explode well before reaching the ground, crushed by the stress of their passage through the very dense atmosphere.
Photograph courtesy of NASA

A radar image of Mercury, showing the intense radar return from one polar cap (top). The other pole is here tipped away from Earth and is not visible. The caps appear to be made of ice, condensed from the vapor of exploding comets and asteroids that have impacted on Mercury's dry, pockmarked surface.
Photograph courtesy of NASA

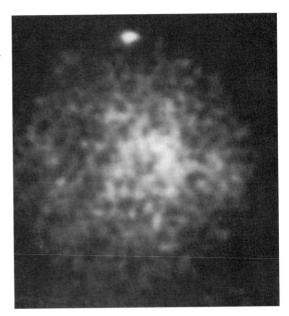

Earth's distant early warning system for protection against cosmic attack, the University of Arizona's Spacewatch camera, produced this image showing the discovery of two new asteroids. The three boxes in each set show the asteroids' locations at the time of three different observations over the course of a few hours. One of the asteroids is a fast-moving near-Earth asteroid, the other a more sedate Belt asteroid.

Photograph courtesy of Tom Gehrels

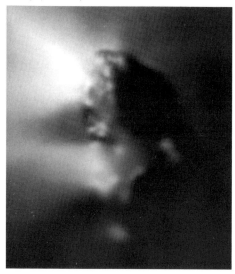

The nucleus of Halley's Comet, photographed by the European Space Agency's *Giotto* spacecraft in 1986. The finest visible detail is on a scale of about 60 meters (200 feet). The very irregular, very black surface dust layer of the nucleus is pierced by several small holes, from which gasses jet outward, boiled off by the Sun's heat from the ice-rich interior.

Photograph courtesy of the European Space Agency (ESA)

The pockmarked, fractured face of the Martian moon Phobos here reminds us of the possible weakness of asteroids that have been bombarded by countless impacts. The parallel "grooves" reflect the drainage of surface regolith into long, deep fractures in the underlying bedrock.

This Viking Orbiter image is courtesy of NASA

past few thousand years. In the last nine thousand years there has been only a single global temperature excursion larger than 0.5 degrees! Human culture seems to have risen in response to an astonishingly improbable stretch of warm, stable climate. Life on Earth is enormously robust. The human race is also very robust. But human culture is disturbingly frail. It is the collapse of civilization—the loss of thousands of years of the fruits of the arts, religion, and the sciences—that we should most fear.

9

EARTH'S TWIN

In striking contrast to the other planets, Venus has a cratering record that is virtually unmodified by subsequent impact, volcanic, tectonic, or eolian processes. However, this record differs from those of other planets because of the dense Venusian atmosphere. On Venus, no craters have evaded Magellan's orbital gaze, for no solitary crater is smaller than 1.5 km in diameter.

GERALD G. SCHABER, 1992

Venus, Earth's aberrant twin sister, should receive almost exactly the same bombardment as Earth. But uncovering Venus's bombardment history has not been easy. A dense, permanent cloud layer of sulfuric acid droplets makes it impossible to see the surface of Venus from space. Why not do what we did for the Moon, Mars, and Mercury, and carry out our mapping by photography and electronic imaging? The answer is simple: the global cloud layer is utterly impenetrable at ultraviolet, visible, and infrared wavelengths. Faint scattered sunlight does reach the surface, as on a very cloudy day on Earth. In fact, several Soviet spacecraft have actually landed on the surface of Venus, survived the hellish conditions of crushing pressures and red heat for several minutes, and transmitted back television images of a few hundred square meters of the surface near their landing sites. Even a giant x-ray machine would be no help: the atmosphere of Venus near the ground is so dense that highly penetrating x-rays would be absorbed after traveling only a few centimeters. Only at radio and TV (and thus radar) wavelengths can we see through the atmosphere all the way to the ground.

Radar studies of other planets first became possible in the early 1960s as a direct result of rapid improvements in radar technology and the construction of moderately large astronomical radar telescopes in Massachusetts (the Millstone radar) and California (Goldstone). When radar was turned on our cloud-shrouded sister planet, it was found very early that the rotation of Venus is astonishingly slow, taking about 243 Earth days. Further, the rotation of Venus was found to be retrograde, in the direction opposite the rotation and orbital motion of most solar system bodies. Its orbit takes it around the Sun in 224.701 Earth days. Its synodic period, the time it takes to lap Earth in its orbital motion, is 583.921 Earth days (1.59866 Earth years). The synodic rotation period of Venus (the time it takes to rotate as seen from Earth) is 145.9277 Earth days. In one orbital synodic period, 2.59866 Venus years, Venus rotates 2.40278 times with respect to the stars, or 4.00144 times relative to Earth. Thus the face of Venus presented to Earth at the time of passage of Venus between Earth and the Sun, when the distance between Venus and Earth is a minimum, is nearly identical to that seen at the previous conjunction. Despite the closeness of this number to 4.00000, the evidence clearly shows that the relationship is not exact: there is no spin-orbit resonance between Venus and Earth.

The same radar data that permit determination of the rotation speed of Venus also permit us to prepare maps of its surface. Radar mapping is carried out by using a very powerful radar transmitter to send bursts of radar power at the target. The signal, spreading out as it races away from the transmitter, drops off in power with the square of the distance it travels through space. After a time delay of several minutes, the time it takes light to travel interplanetary distances, the radar pulse hits the nearest point on the surface of Venus, which is the point at which Earth is located directly overhead, at the zenith. (Reflections from more distant parts of the surface, of course, take longer to get there and longer to return.) The reflected signal begins its long trip back to Earth, also dropping off in power with the square of the distance it travels. Overall, the intensity of the returned signal drops off with the fourth power of the distance of the target from Earth. To make the best of the very weak returned signal, the receiving antenna on Earth is chosen to be the largest and most sensitive available, and need not be the same one that sends the signal in the first place.

Venus at its point of closest approach (inferior conjunction) is about 0.3 astronomical units away from Earth. When Venus and Earth are on opposite sides of the Sun (and the Sun is in the way!) the distance is about 1.7 AU. The strength of the returned signal drops off from closest to farthest distance not as 1.7/0.3 (a factor of about 5.7), but as the fourth power of this ratio, 5.7 x 5.7 x 5.7 x 5.7 = 1056. Thus radar mapping Venus near closest approach is some one thousand times faster than when Venus is at the opposite side of its orbit.

Because of this sensitivity problem, and because of the near-resonant rotation of Venus, we acquire superb, detailed maps of Venus only near the time of conjunction—but each successive conjunction maps almost exactly the same terrain we saw last time! True, the lack of perfect resonance allows Venus to drift by 0.00144 rotations (about half a degree of longitude) each time it passes by. To get global coverage of the equatorial regions with the highest sensitivity we only need to wait and observe Venus at 695 consecutive conjunctions. Unfortunately, that requires an observing program lasting 695 synodic periods, or 1111.3 years!

This practical problem is compounded by the fact that we cannot clearly see features that are located far from the equator. Early on, studies of the equatorial regions revealed great volcanic mountain systems and several apparent large impact craters. However, Earth-based observations of the polar regions of Venus require us to observe nearly horizontally through a dense, absorbing atmosphere. Almost no reflected signal gets back to Earth, making high-quality polar mapping impossible for Earth-based radar.

How, then, do we explore the geology of our twin sister planet? Trying to map the planet with landers a few hundred square meters at a time is totally crazy. But there is a way to map Venus: if Earth-based radar and lander photography won't do the trick, what about a small, sensitive radar close to Venus? The idea makes both technical and economic sense. In the 1970s, NASA advisory committees studied the possibility of building a Venus Radar Mapper (VRM), sometimes called the Venus Orbiting Imaging Radar (VOIR) after the French verb "to see."

Building upon NASA experience with Earth-orbiting radars, both NASA and the U.S. Air Force planned missions using space-based radars. As early as 1974, Soviet military satellites with nuclear-reactor-powered radars were placed in orbit several hundred kilometers above the surface of the Earth to track American carrier task forces, so the principle was already demonstrated in practice. NASA's designs evolved with time into the *Magellan* Venus Radar Mapper. At the same time as the early NASA studies, the air force settled upon a spacecraft code-named *Lacrosse*. This satellite weighs in at twenty tons, so large that only the space shuttle or the *Titan 4* could lift it into low-altitude, high-inclination orbit around the Earth. The task of the *Lacrosse* satellites was to use radar to penetrate cloud cover and monitor the deployment of surface weapons systems such as mobile missile launchers and tanks. The first *Lacrosse*, launched by the space shuttle in December of 1988, was placed in an orbit with an inclination of only fifty-seven degrees to the equator, affording good coverage of Eastern Europe and most of the Soviet Union. Its most important function was to monitor Iraqi armored vehicles during the Gulf War.

The Soviet Union, building upon the technological base of its military ocean-surveillance radar satellites, built two radar mappers for missions to Venus, the *Venera 15* and *16* spacecraft. Both were launched by separate Proton boosters in June of 1983. Their radar survey of Venus was of high technical quality but limited in coverage. The presence of large impact craters was confirmed. The appetite for *Magellan*'s global coverage was whetted.

The 975-kilogram *Magellan* spacecraft was launched in May of 1989 and arrived at Venus in August of 1990. It was placed in an eccentric, high-inclination orbit around Venus by the firing of a small rocket engine. Mapping began on September 15, 1990. Over the next few months, *Magellan* successfully mapped over 90 percent of the surface of Venus with much higher resolution than was possible from Earth.

Magellan mapped a tremendous variety of geological features, including many hundreds of impact craters, during its first 243 days of mapping (one complete rotation of the planet beneath the spacecraft's orbit). Interpretation of the vast library of information in these radar images continues, but a number of firm conclusions are already evident.

First, the surface of Venus is not extremely ancient. Based on the frequency of large craters and their distribution over the planet, it appears that Venus was completely resurfaced in a geologically short interval of time (probably less than 100 million years) only a few hundred million years ago. At that time, all evidence of older features was erased. All the craters we see simply reflect recent solar system bombardment processes.

The size distribution of craters on Venus is very different from anything seen on the Moon, Mars, or Earth. Craters with diameters of twenty to one hundred kilometers are most numerous. On the Moon and Mars, by comparison, the frequency of craters increases steadily toward smaller sizes; each size crater is always more abundant than the next larger size. Earth displays an intermediate behavior, with craters less than a kilometer in diameter being rarer than larger ones. On Earth, an easy explanation suggests itself: the active erosion at Earth's surface tends to erase or fill small impact scars quite quickly, making them much harder (or even impossible) to find. But reflection upon the Tunguska event has led us to a second factor, the breakup of weak, fast-moving bodies as they transit the atmosphere of either Venus or Earth.

Venus differs from Earth in two very important ways as a target for comets and asteroids. First, because Venus is closer to the Sun and deeper in the Sun's gravity field, orbital velocities are higher there than near Earth. Second, the atmospheric pressure reaches over ninety-two times that at Earth's surface, and the density of the atmosphere at the surface of Venus is fifty-six times as high as on Earth. Since the aerodynamic pressure at the front of an entering body is proportional to the density of the atmosphere and to the square of its velocity, it is easy to show that the dynamic pressure on a body deep in the atmosphere of Venus is roughly one hundred times as high as at a comparable altitude on Earth. Thus an entering asteroidal or cometary fragment that barely has enough strength to survive to the surface of Earth would be crushed by the dynamic pressure of passage through the atmosphere of Venus while still high above the surface. One might also wonder whether the high temperatures in the lower atmosphere of Venus would assist in melting and breaking up entering bodies. The answer is that the entry process takes only a few seconds, far too short a time for heat to be conducted into the interior of the entering body.

The rarity of middle-sized craters on Venus goes hand in hand with an even more startling observation: there are *no* craters on Venus smaller than 1.5 kilometers in diameter! Further, there are important differences in the appearance of craters of different sizes. Large (40- to 280-kilometer) craters look like the large double-ringed and multi-ringed basins on Mercury and the Moon. The 15- to 50-kilometer craters have the familiar central-peak structures seen elsewhere in the solar system. There are eight craters with diameters greater than 100 km that have formed on Venus within the past (roughly) 500 million years. On Earth, the only known craters larger than 100 kilometers and less than 500 million years old are Manicouagan, Quebec (100 km; 212 million years), Popigai, Siberia (100 km; 35 million years) and Chicxulub, Yucatán (>180 km; 65 million years). Allowing for the fact that less than a quarter of the present surface of Earth is more than 200 million years old, for the possible differences between the asteroid and comet collision rates at Earth and at Venus (and implicitly for the uncertainties in the age of the resurfacing event on Venus calculated from these uncertain fluxes), and also allowing for the somewhat higher impact velocities at Venus, these numbers are in tolerable agreement. On Venus, there are ninety-two craters with diameters of 40 to 99 kilometers, compared to eight of similar size and age on Earth. The story of the large craters on Venus and Earth is consistent if roughly 60 percent of such craters on the continents of Earth are yet to be discovered, or have been destroyed by erosion. In view of the intense and continuing geological activity of Earth, that is not an unreasonable assumption.

However, most craters smaller than 15 kilometers diameter on Venus are markedly noncircular in outline and irregular in appearance. Craters smaller than about 3 km diameter are mostly multiple, with two to a dozen or more distinct impact craters of irregular outline closely clustered together. Here we see very clearly and graphically the results of the breakup of small impactors in the atmosphere.

Multiple craters are not unique to Venus, but those seen on other planets tend to be composed of exactly two craters of the same age, located several diameters apart. Most of the multiples on Venus are both more compact and more complex than these.

Another remarkable feature of impact processes on Venus is that the smaller craters are often surrounded by broad circular or parabolic expanses of radar-dark (smooth) materials. When the craters depart most from circularity, the dark area around the crater tends to be parabolic in outline. Ouside the dark region the surface is slightly brighter than the background terrain. There is a smooth sequence from these small multiple craters to dark circles with bright outer rings, with almost no detectable crater at the center. These in turn merge gradually into dark circles and bright outer rings with no evidence of impact cratering. At the extreme, one finds a tiny, faint dark region surrounded by a broad, diffuse collar of brightened ground. Evidently we are seeing the effects of a range of breakup behaviors. The craterless features are made by entering bodies that explode and fragment into tiny bodies, too small to excavate craters, well above the planetary surface. Each aerial explosion lays down a powerful shock wave, much more intense than one from a similar-sized explosion at the same altitude on Earth. The shocked gas in the blast wave near the surface of Venus can actually have a density close to that of liquid water! The blast pressure can reach one thousand atmospheres (one metric ton per square centimeter; fifteen thousand pounds per square inch). Directly beneath the explosion the blast wave strikes the surface vertically, heating it strongly and compacting it, but not scouring effectively. Intense radiative heating of the ground by the heat of the fireball can melt a thin layer of surface materials, and the blast wave, arriving after the peak radiative heating, can scour the ground near the explosion epicenter and deposit that loose dust and dirt in a ring farther out.

Many craters on Venus show evidence of oblique impacts, as deduced for the Tunguska body on Earth, and directly demonstrated by the Campo del Cielo and Pampas crater chains in Argentina. Because of the absence of water erosion and the low wind speeds on Venus, post-impact weathering of these craters is not obvious.

Suppose that half the large impactors on Venus are comets, and that half the "asteroidal" bodies are extinct comet cores containing abundant, deeply buried ice. Then some 50 percent of the mass of material falling onto Venus is water. We know from spacecraft studies of the atmosphere that Venus not only is far too hot to have liquid water on its surface, but also has a very dry atmosphere, containing only about two grams of water per square centimeter of surface area. If fully condensed out of the atmosphere, there would be enough liquid water to cover the surface in a layer just a little less than one inch deep! On Earth, the amount of water in the oceans and atmosphere is one hundred thousand times as large. It is very difficult to see why two adjacent planets should accrete such radically different materials.

Is it possible that the two planets started out similar to each other and followed divergent evolutionary paths? In other words, should we interpret the tiny amount of water on Venus as the last residue of once-mighty oceans that have been escaping for billions of years?

The best answer that we can presently offer is that an impact capable of producing a 280-kilometer diameter crater on Venus (Mead crater, named after anthropologist Margaret Mead) would require a cometary impactor with a diameter of at least 28 kilometers, bearing more water than the entire present inventory on Venus. The total water supply from all the impactors whose craters were seen by *Magellan* suggest the infall of several times the present water inventory over the last few hundred million years. On the average, it would take a few tens of millions of years for infall of comets and asteroids to provide the amount of water now present on Venus. Further, the present rate of escape of hydrogen from Venus is enough to exhaust all the water now present in about 30 million years, less than 1 percent of the age of the solar system. There is therefore every reason to believe that the water content of the atmosphere of Venus is dominated by recent events, and preserves no record of the ancient history of the planet.

It is interesting to apply the same reasoning to Earth. The comet and asteroid influx of water to Earth should be about one centimeter per ten million years. The escape rate of hydrogen from Earth is low enough so as to be unimportant by comparison. The present water content of the oceans (a globally averaged water layer three kilometers deep) would require 3,000 billion years to accumulate at present rates, but Earth is only 4.6 billion years old. Of course, we know from the cratering history of the Moon that the impact rate was much higher at times earlier than 4 billion years ago, during the era when the planets were still sweeping up vast amounts of interplanetary debris and closely approaching their present mass. It is not impossible that impacts in ancient times brought in Earth's supply of water, but it is extremely unlikely that a significant fraction of Earth's abundant water supply was brought in more recently than 4 billion years ago.

What would be the effects of such an impact history on Mars? The very low atmospheric pressure would provide only a feeble barrier to the passage of impactors and the expansion of explosion shock waves. Also, the low escape velocity makes it much easier for explosion debris to escape from Mars than from either Venus or Earth. Hampton Watkins and I suggested in 1982 that the present-day bombardment of Mars by comets and asteroids would have a net erosive effect on the atmosphere: over long spans of time, the total mass of the Martian atmosphere may have been reduced by a factor of one hundred by impact erosion. More recent and much more sophisticated calculations of the behavior of large explosions by Ann Vickery and Jay Melosh confirm that impacts can have a powerful erosive effect on the atmosphere of Mars.

The experience of studying Venus with the *Magellan* radar mapper has given us truly enlightening insights into the phenomenon of breakup of entering projectiles. Increased knowledge of the behavior of airbursts has led to a better understanding of mysterious aerial explosions that occur frequently on Earth. It has become increasingly clear that the Tunguska explosion was just the most spectacular recent example of a class of phenomena that are very important, but also very elusive. Many aerial explosions that cause devastating damage over areas of hundreds to thousands of square kilometers may leave no discernible evidence a few decades after their occurrence.

Photographic networks devoted to tracking fireballs are an inadequate means to investigate kiloton to megaton explosions because these networks cover areas of only a few hundred thousand square kilometers, compared to the roughly 500 million square kilometers of surface area on the planet. Suppose that four hundred thousand square kilometers is covered by a photographic network that operates with dark, clear skies an average of six hours per day. This is equivalent to one hundred thousand square kilometers of continuous coverage. Suppose we need data on the frequency and breakup behavior of six-meter asteroidal fragments with explosive powers of ten kilotons (slightly less than the Hiroshima or Nagasaki explosions). According to preliminary Spacewatch data, about ten such objects should strike Earth each year, or one per 50 million square kilometers per year. The probability that our network will detect an object this large in its first year of operation is then 1 in 500. On the average, the network will have to run for one thousand years to have a 75 percent chance of observing a single body of that size and explosive power. This is not a high rate of return on our investment!

However, unknown to most people, we *have* been investing in a highly capable fireball detection and tracking system for many years. The Defense Support Program (DSP-647) satellites, which operate in geosynchronous orbit thirty-five thousand kilometers above the equator, carry infrared telescopes with sensitive imaging detectors designed to detect and track the heat from the engines of military ballistic missiles. They provide near-global coverage (the degree of coverage changes with time and is in any case classified) with the sensitivity to track bombers, civil aircraft, trains, and fireballs in addition to ICBMs, IRBMs, and SLBMs. An outgrowth of technology developed in the low-Earth-orbit Midas satellite series in the 1960s, the first geosynchronous-orbit ballistic missile early warning system (BMEWS) payloads were launched under Air Force Program AFP-949 in 1968 through 1970. The first DSP-647 satellites were launched in November of 1970 and May of 1971 under the IMEWS (infared missile early warning satellite) label. Technological upgrading has continued over the years, with the newest series of DSP spacecraft initiated in 1989.

The public release of data from sensitive military intelligence-gathering systems was unthinkable until the collapse of the Soviet empire. But now coy peepshows of DSP data are conducted even before audiences containing top Russian weapons experts. The game is to release information useful to the scientific community without compromising information revealing the sensitivity and operational procedures of the system. The bottom line is that the DSP satellites find that there are about a dozen ten- to twenty-kiloton explosions in the atmosphere per year, in good agreement with the Spacewatch data. Between 1975 and 1992, the DSP system observed 136 large fireball entries, not one of which reached the surface of Earth intact. Unfortunately some valuable data have been lost: changing DSP management practices over the years has led to incomplete retention of fireball data from certain eras. Nor has there been any public account of the method of calculating the masses and sizes of entering bodies from their heat emission during entry. Further American glasnost may well clarify these matters in the near future.

Yet another American program for monitoring nuclear weapons tests has deployed ultrasensitive air-pressure sensors called microbarographs on the roofs of numerous American embassies and consulates around the world to detect the pressure waves of nuclear explosions. The data have not yet been made public; however, it is known that this network observed a major event on August 3, 1963, over the ocean between South Africa and Antarctica. This explosion, which reportedly had a yield of at least five hundred kilotons of TNT, was initially misidentified as a secret nuclear weapons test by South Africa or Israel; however, no radioactivity was detected after the blast, and the nuclear explanation is clearly wrong.

Our final source of information about large airbursts is eyewitness reports. It is good to recall that astronomers cover only a tiny fraction of the sky at any time. Most astronomers are enclosed in observing cages or in control rooms for most of the time they are observing. Their telescopes have exceptional sensitivity, but view only tiny areas of the sky. The little wide-angle Schmidt telescopes used for hunting comets cover an area of the sky about the size of the bowl of the Big Dipper. All larger telescopes cover far smaller areas. If a brilliant fireball were to pass overhead in the midst of an observing run, chances are that it would not be seen by the astronomers. But millions of people are out at night. Perhaps 10 percent of the surface area of the planet is dark, clear, and populated at any given time. These untrained observers should witness a ten-kiloton explosion somewhere on Earth about once a year. But what do they see? Our challenge is to collect and correlate large numbers of eyewitness reports of widely varying detail and accuracy in order to determine what really happened. Of course, many witnessed explosions are seen by only one or two people, and there is usually insufficient evidence to conclude very much about the height, velocity, and explosive yield. The best places to get multiple reports are densely populated areas, which means cities. Urban environments imply low air quality, very high levels of light pollution, and serious obscuration of the sky by tall buildings. The challenges are great.

But despite these problems, astonishing numbers of airbursts are observed, complete with brilliant flares of light, explosive disruption at high altitudes, and shock waves affecting large areas on the ground. Consider a few examples:

Northeastern United States and Eastern Canada, March 9, 1822. *A fireball the brightness of the full Moon was seen in Quebec City, Montreal, Boston, and in parts of New Hampshire, Rhode Island, Pennsylvania, Vermont, and Maine. Two powerful explosions were heard from Portland, Maine to Albany, New York. The blasts, which sensibly affected several houses, were followed locally by a strong sulfurous smell.*

—American Journal of Science, *1823*

Council Bluffs, Iowa, November 28, 1894. *A large meteor fell, striking the earth in the southwestern part of the city, about 11 o'clock last night. . . . The most strange phenomenon connected with it is that about two minutes after the meteor fell there was a terrific shock, scarcely less severe than an earthquake, which shook nearly every building in the city and awakened nearly all of the slumbering inhabitants. Buildings in the north part of the town, fully one mile away from where the meteor fell, were violently shaken.*

—The New York Times, *November 29, 1894*

Meadville, Pennsylvania, August 12, 1904. *A meteor taking a northerly direction struck the earth somewhere near Concord station, thirty-five miles east of this city, on the line of the Erie Railroad, a few minutes after 2 o'clock this morning.*

A terrific explosion accompanied the compact with the earth, followed by a high wind lasting fifteen seconds. . . . The explosion shook buildings in Titusville, twenty miles away.

—The New York Times, *August 13, 1904*

Southern New Jersey, April 23, 1922. *A great ball of fire, trailing an iridescent tail like a comet rushed across the sky near the Southern New Jersey coast at 9 o'clock last night and disappeared earthward with an explosion that was heard over a thirty-mile area. . . . All insisted that the lightning-like illumination which accompanied its swift passage across the heavens and the terrific detonation with which it struck, rocking buildings and shattering windows, precluded it having been anything smaller than a heavenly body.*

From Toms River came reports of windows shattered, while it was said that the bursting somewhere in the distance of the strange visitor let loose clouds of noxious gas, which so polluted the atmosphere that persons in the streets held moistened hankerchiefs to their nostrils for fifteen minutes. . . .

The phenomenon lasted more than thirty seconds, in the course of which they saw the fire ball plunge earthward and felt it strike somewhere and rock the earth. The explosion they said was deafening.

—The New York Times, *April 24, 1922*

Wynard, Saskatchewan, July 25, 1922. *Residents of this section today were swapping accounts of their experiences yesterday, when a meteor fell in Big Quill with a great roar and explosion that frightened people and animals for miles around.*

—The New York Times, *July 26, 1922*

Portland, Maine, July 18, 1926. *A crash which Professor Charles Hutchins of Bowdoin believes to have been the bursting of a meteor awakened thousands of people in Maine today at 5:08* A.M. *Daylight Time. This city, Dexter, 130 miles north, and Saco, 15 miles south, are among the places reporting the phenomenon.*

The crash was preceded by a blinding bluish light.

—The New York Times, *July 19, 1926*

Atlanta, Georgia, May 23, 1928. *Flashing through the skies just after midnight Wednesday, a huge meteor raced over South Georgia and a part of South Carolina, exploding in midair with a shock that frightened thousands and caused people of many cities to believe that it had brought an earthquake in its trail.*

—The New York Times, *May 24, 1928*

Ribinsk, Russia. *A meteor burst with tremendous noise in this city, spitting fire in all directions.*

—The New York Times, *November 17, 1929*

Chita, Siberia, January 26, 1930. *Reports reached here today that a huge meteorite with a thunderous roar and a dazzling light that extended for many miles fell into the woods near the Mongolian border yesterday.*

—The New York Times, *January 27, 1930*

Malinta, Ohio, June 10, 1931. SHOCK IN OHIO LAID TO A GIANT METEOR. *A terrific shock jarred six Ohio counties today, rocked houses in dozens of towns in the State and Indiana and roused thousands of sleeping persons. . . . Lynn Fredericks, a telephone lineman living at Napoleon . . . said he smelled sulfur immediately after the blast.*

W. K. Gunter of Malinta, one of the first to reach the hole, said that so great was the force of the blast that a nearby tree had been tossed several hundred feet. . . . In Findlay, twenty-five miles away, the shock was so distinct that objects were rattled off the desks at the police station.

—The New York Times, *June 11, 1931*

Amarillo, Texas, March 24, 1933. *A gigantic meteor lighted the skies with awesome brilliance in five Southwestern States before dawn today and, with a thunderous rumble that rattled doors and windows, apparently disintegrated on its earthward plunge.*

—The New York Times, *March 25, 1933*

Calgary, Alberta, March 18, 1934. *A meteor which flashed into view here tonight for an instant is believed to have exploded somewhere in Central Alberta with such force that houses in Trochu, Delburne, Irricana and Bashaw were shaken and terrified residents of Bashaw rushed from their homes.*

—The New York Times, *March 19, 1934*

San Francisco, California, September 28, 1934. *Plowing head-on at an altitude of 7,000 feet into a great shower of meteors, some of which exploded with sufficient force to rock the ship, a crowded New York–San Francisco plane of the United Air Lines figured in a sensational sky trip at dawn today, which gave crew and passengers the experience of a lifetime.*

—The New York Times, *September 29, 1934*

Krasnovishersk, U.S.S.R., July 8, 1935. *A meteor that passed over this town in the Ural area in a fiery streak today exploded with such force that it shook houses throughout the district. Its passage, high over the town, was accompanied by a thunderous noise. The explosion occurred presumably before the meteor struck the earth and while the sky was still marked by a smoky trail.*

—The New York Times, *July 19, 1935*

St. John's, Newfoundland, October 19, 1936. METEOR SHOWER SETS SKIES AFLAME. *Newfoundland Sees Balls of Fire Exploding and Strik-*

ing Sea—World's End Feared. . . . Burnt Island, on Placentia Bay, reported a large flame and heavy explosion in a northerly direction. . . . So brilliant was the illumination that, even in midday, people rushed to windows in alarm.

—The New York Times, *October 20, 1936*

Portland, Oregon, July 2, 1939. EXPLOSION IN SKY ROCKS PORTLAND, ORE., AFTER METEOR "BIG AS MOON" IS SIGHTED. *A mysterious explosion which, observers believe, was caused by a huge meteor, rocked Portland and the surrounding territory just before 8 A.M. (Pacific standard time) today. Scores of persons reported that they looked skyward in time to see a vast burst of smoke and spurting flame. The shock was felt over a radius of forty miles.*

—The New York Times, *July 3, 1939*

Philadelphia, Pennsylvania, May 4, 1945. METEOR BLAST ALARMS RESIDENTS ALONG THE MID-ATLANTIC SEABOARD. *Hundreds of thousands of persons in eastern Pennsylvania, New Jersey, Maryland and Delaware were awakened early today by a blue-white flash and a series of explosions and tremors. The light streaked across the sky at 3:38 A.M. and the subsequent detonations shook houses, rattled windows, burst open doors and sent thousands of startled persons rushing to telephones with a deluge of inquiries. . . . Most of those who actually observed the phenomenon described the following blast as a major explosion; some saying it resembled a fierce wind.*

—The New York Times, *May 5, 1945*

London, England, November 30, 1946. *A flash that lighted a large part of the Forest of Dean and an explosion that rocked the nearby town of Coleford, Gloucestershire, last night has puzzled Air Ministry weather experts. The explosion plunged the town into darkness, wrecked telephones, knocked a boy off his bicycle, blew a torch from a woman's hand and set bottles dancing in taverns. Some believed it was caused by a meteorite. . . .*

—The New York Times, *November 30, 1946*

Elizabeth, N.J., March 14, 1948. JERSEY EXPLOSION PROVES A MYSTERY. *Some persons reported that they had heard a loud report, while others said they felt only the impact of an explosion. . . . No earthquake or other disturbance was recorded on seismograph instruments. . . . Officials*

· ·

pointed out that an "air explosion," that is a disturbance not having its origin in the earth, would not be recorded on the instruments.
—The New York Times, *March 15, 1948*

Glen Cove, Long Island, October 18, 1952. TWO MYSTERY BLASTS ROCK LONG ISLAND. *The north shore of Long Island—from Glen Head to Cold Spring Harbor—had a mystery on its hands today that stumped everybody. . . . At 11:16 A.M. two blasts rocketed from the skies. Seconds after the concussion, there was hardly a person left indoors.*
—The New York Times, *October 19, 1952*

The list of spectacular aerial explosions runs on and on, for hundreds of entries, beginning with the earliest periodical publications. The frequency of published reports has actually *declined* since 1960 because people tend to dismiss loud explosions as merely military sonic booms. Earth's atmosphere, though not as effective a shield as Venus's denser blanket of gases, nonetheless suffices to destroy vast numbers of bodies which otherwise would have devastating effects on the ground. The testimony of Venus is that aerial fragmentation of projectiles is extremely important. The cratering behavior seen on Venus adds another valuable perspective to our understanding of the bombardment of Earth.

10

YOU FOUND *WHAT* ON MERCURY??

Icy surfaces are common in the Solar System, particularly on the moons of Jupiter and the rest of the outer planets. However, we were surprised to find ice could exist on Mercury, the planet nearest the Sun. Temperatures on Mercury reach as high as 700 K at the equator. Nevertheless, we interpret that a highly radar-reflective region observed on the north pole of Mercury in August 1991 is due to ices.

MARTIN SLADE, BRYAN BUTLER, AND DUANE MUHLEMAN
Science, *October 1992*

The aridity of the Sahara Desert is legendary. The redness of Mars has been universally known for millennia. The desolation of the Moon is a commonplace to astronauts, poets, and scientists alike; the concept may even be familiar to university administrators. The scorched, barren wasteland of Mercury, utterly devoid of any trace of air or water, is a touchstone of the space age, advertised and memorialized by television specials and cereal boxes.

Oh well, three out of four ain't bad.

Mapping Mercury has been a challenge from the start. It orbits closer to the fires of the Sun than any other planet, well inside the orbit of Venus. Not only is Mercury much smaller than Venus, but at its closest it is twice as far away from us as Venus at its best. Of course, when Mercury is at its closest to Earth it is at inferior conjunction, directly between Earth and the Sun, and quite invisible in the blinding glare of our favorite star. The best time to observe Mercury through terrestrial telescopes is when it is at its greatest angular distance from the Sun (elongation), quite a bit farther away from Earth than at conjunction, and much smaller in apparent size. But even at elongation, Mercury never gets more than about 27° from the Sun. It is sometimes briefly visible to the naked eye low in the western sky immediately after sunset or in the eastern sky just before dawn. Mercury can only be seen in a dark sky when it is nearly on the horizon, severely distorted by the lengthy trip through Earth's dusty and turbulent atmosphere. So disconnected are these rare observations that the Romans thought they were two different objects, called by them Mercury and Apollo, respectively. Copernicus is said to have never seen Mercury at all.

For centuries, Earth's greatest astronomers struggled to map the faint surface features of Mercury, seen through long path lengths of Earth's atmosphere, often in an imperfectly dark sky. Schiaparelli reported faint canal-like features similar to those he saw on Mars. The American astronomer E. E. Barnard, like most observers, saw not *canali* but faint nonlinear markings, rather like "those seen on the Moon with the naked eye." Percival Lowell, another experienced canal-watcher, saw linear markings. Whatever markings were seen, they did not shift detectably across the face of Mercury in the course of a night. It soon became widely accepted that Mercury, like the Moon facing Earth, always kept the same side facing the Sun. The rotation period and the orbital period (Mercury's day and year) must be the same, a situation called a 1:1 spin-orbit resonance. The supposed permanent subsolar point would be the hottest place in the solar system outside the Sun. The permanent antisolar point would be extremely cold.

When radar was directed toward Mercury in the early 1960s the situation changed overnight. It was immediately found that Mercury rotates one and a half times per orbit. Thus, at consecutive perihelion passages, opposite points on the equator pointed at the Sun. Mercury's eccentric orbit causes the intensity of sunlight on its surface to be twice as high at perihelion as at aphelion, so these two alternating points get very strongly heated at perihelion compared with the rest of the planet. Mercury therefore has two "hot poles" on opposite sides of its equator. The rest of the equator does not get quite as hot. The coldest points on the planet would be the poles, or, more precisely, permanently shadowed low points close to the poles.

Radar mapping of Mercury reveals a Moonlike cratered surface and no evidence of an atmosphere or oceans. Spectroscopic studies from Earth-based telescopes had found no trace of an atmosphere, reinforcing the view of Mercury as a planet devoid of volatile materials.

As we discussed in more detail in chapter 5, the *Mariner 10* mission in 1974 used a Venus close swingby to deflect the spacecraft inward toward Mercury. *Mariner 10* achieved an orbit that grazed Mercury's orbit at perihelion, but reached out so much farther from the Sun

. . . .

at aphelion that it had an orbital period exactly twice the length of Mercury's year. This resonant condition permitted *Mariner 10* to fly by Mercury at close range at the times of *alternate* perihelion passages. Thus the portion of Mercury illuminated by the Sun was exactly the same each time *Mariner 10* flew by: half the planet could not be photographed because it was in darkness every time the spacecraft passed near enough to photograph the surface.

Mariner 10 carried out three successful close flybys of Mercury over a period of four Mercury years (six Mercury rotations), transmitting stunning TV images of the heavily cratered surface back to Earth. Like the Moon, Mercury is dominated by ancient, very heavily cratered, highland terrain. Images of the polar regions reveal many large, deep craters near the poles, providing opportunities for permanently shadowed crater floors and very low local temperatures. But the incredibly tenuous atmosphere, as rarified as the best vacuum attainable on Earth, lacks detectable traces of any material that might condense at these low temperatures. Indeed, the principal components of the atmosphere were found to be atomic hydrogen, a transient trace species temporarily trapped out of the solar wind, and atomic sodium, produced perhaps by solar baking of the surface near the hot poles or by tiny high-velocity meteorites which vaporize upon impact with the surface.

In 1991 a team of radar astronomers from the Jet Propulsion Laboratory and California Institute of Technology, Martin Slade, Bryan Butler, and Duane Muhleman, embarked on an observing run designed to provide the best-ever map of the surface of Mercury. The basic technique of radar mapping of another planet is to bounce a short, sharp radar pulse off the planet and then dissect the reflected signal. The wave front of the pulse strikes the closest point on Mercury first, and the reflected pulse from that point begins the long trip back to Earth well before the transmitted pulse even reaches the more remote areas of Mercury's surface. The first signal to arrive back at Earth is the mirror-like reflection of the small area about the sub-Earth point on Mercury. The reflected pulse from rings of terrain farther from the sub-Earth point (and farther from Earth) arrive at progressively later times, ending with the signal reflected from the edges of the planetary disk as seen from Earth. The edge of the visible disk reflects mostly in the forward direction, so that only a weak signal gets back to Earth. Thus the delay time of the returned signal tells us about the range of the reflecting surface from the

. . . .

transmitter. Each interval of delay time defines a ring of terrain about the sub-Earth point. All points in a given delay ring are indistinguishable from one another using this information alone.

Fortunately, there are other kinds of information carried by the reflected pulse. For example, the rotation of Mercury and the angular rate of apparent rotation of Mercury due to the relative orbital motions of Mercury and Earth cause different parts of the surface to have different radial velocities relative to Earth. Regions of the surface that are approaching Earth at the time of arrival of the incident pulse reflect a signal with a slightly higher frequency than that of the arriving pulse, and regions retreating from Earth will reflect a signal that is Doppler-shifted in the opposite sense, to lower frequencies. The band on the surface of Mercury from pole to pole through the sub-Earth point along which the surface is neither approaching nor receding from Earth relative to the center of Mercury (the sub-Earth meridian) contains the mirror-reflecting sub-Earth point, and gives the brightest return. The visible disk of the planet is divided into slices parallel to this band, with each slice Doppler-shifted by an amount proportional to its distance from the sub-Earth meridian, but with opposite signs (frequency increase vs. frequency decrease) depending on the local line-of-sight velocity (approaching vs. retreating). Thus each feature on the surface of the planet is characterized by a particular combination of delay and Doppler shift. The process of constructing a map by use of this information is called delay-Doppler mapping. This is exactly how both Venus and Mercury were first mapped by radar.

The main problem with constructing such a map is that there are circumstances in which the map is highly ambiguous. Consider observing a planet when the sub-Earth point is on the planet's equator. At that time, the reflected signals from two points at the same longitude and equal but opposite latitudes (say, 30° N and 30° S) will have both the same Doppler shift and the same delay. This causes a complete north-south ambiguity of the map: there is no way to distinguish northern from southern features.

Fortunately, the orbit of Mercury is inclined by seven degrees relative to the plane of Earth's orbit. This means that by observing at different times, we can see over one pole or the other, and use the different orbital motions to help disentangle this ambiguity.

Another kind of information contained in the reflected signal is its polarization. We are all familiar with the behavior of polarized electromagnetic radiation from Polaroid sunglasses. The vibrations of the radiation transmitted from the radar telescope may be linearly polarized or circularly (corkscrew) polarized. When a circularly polarized signal is reflected from a flat surface, the direction of polarization is reversed. When reflected from a sawtooth surface, each wave is reflected (and reversed) exactly twice, and emerges with the same polarization direction as the incoming wave. When reflected from a rough surface, singly and doubly reflected waves intermingle, and the signal is depolarized. But observations of other solar system bodies with icy surfaces have a peculiarly high ratio of same-sense to opposite-sense polarization of the reflected signal. The explanation of this fact is subtle, but the correlation with observations is very strong; indeed, no exceptions are known.

Slade and company chose to sidestep the limitations of mapping a small and distant disk by a technique that requires splitting up the total returned signal into a very large number of "bins" with different frequencies and time delays, as is done in delay-Doppler mapping. To make their alternative mapping scheme work, and to achieve the best possible resolution of the surface of Mercury, they needed the most sensitive (largest-area) receiver on Earth. They therefore transmitted their pulses from the Goldstone antenna in the Mojave Desert, but received the returned signal at the Very Large Array radio telescopes near Socorro, New Mexico. They used a technique known as aperture synthesis to produce the map, in effect electronically coupling the signals received by the array of radio telescopes so as to simulate the effect of a single giant ultra-high-resolution receiver.

The results were wholly unexpected: Mercury has small, reflective polar caps with the distinctive depolarization behavior of ice. These polar deposits are found partly covering the surface in two regions that correspond to large, deep craters near the two poles. The ices are most likely buried meters deep beneath a layer of surface dust. The temperatures of these permanently shadowed crater floors are so low that evaporation of the ice is incredibly slow: a layer of ice meters thick would last for many billions of years, as long as the age of the solar system. However, ice buried meters deep may be excavated at any time by a small impact explosion. Impacts into the ice layer would eject crushed permafrost onto the surface, vaporize the nearby ice, and produce very hot water vapor that would have little difficulty escaping from the planet.

Where might this ice come from? It is one thing to say that ice, once present, could persist for very long periods of time without evaporating away. It is quite another to say that there is a source of water on the most thoroughly baked planet in our solar system! The answer appears to lie in the same impacts that remove ice. Both comets and water-bearing asteroids impact on Mercury, as on Venus and Earth. Comets are very rich in water, but they impact on Mercury at extremely high speeds, sometimes surpassing one hundred kilometers per second. So energetic is the impact explosion that it not only removes all the water in the impactor, but a fair amount of the surface of Mercury along with it. But water-bearing asteroids are a different matter. These bodies, which circle the Sun in more nearly circular orbits, can approach Mercury at much lower speeds than the typical comet can. In addition, the gases from asteroid impact explosions contain much higher proportions of rocky materials, which are heavy and slow down the fireball expansion speed. A significant fraction of their water content can emerge from the explosion at a speed below the escape velocity of Mercury. These low-speed water molecules bounce around randomly on the surface until they hit a region hot enough to kick them off the planet, or until solar radiation dissociates or ionizes them—or until they hit a surface so cold that they stick to it. Then ice forms.

This astonishing discovery of polar ice on Mercury makes it clear that impacts play a major role on all the terrestrial planets. We saw in chapter 9 that impacts on Venus have the effect of adding water and other volatile materials to its atmosphere, compensating for destruction of water by ultraviolet sunlight and the subsequent loss of hydrogen from the planet. We saw in chapters 5 and 7 that impacts have profoundly shaped both the surface and atmosphere of Mars, causing the erosive loss of the large majority of the early atmosphere as a direct result of impact explosions that accelerate atmospheric gases to speeds well above the planetary escape velocity. Even quite small bodies can penetrate the tenuous atmosphere of Mars and cause erosive explosions on its surface. In chapter 2 we saw that these large impacts also occur on Earth, and in chapter 8 we encountered a lengthy list of devastating consequences to the surface, atmosphere, oceans, and life. Perhaps the central phenomena of large impacts on Earth are the raising of vast dust clouds and the production of nitrogen oxides by the blast wave. Significant loss of atmosphere from Earth by explosive blowoff can occur in only a tiny proportion of all impacts. And finally, we have seen in this chapter that impacts both supply water to Mercury and limit its accumulation there. Thus the four terrestrial planets are all affected in important but very different ways by comet and asteroid impacts. The only common feature is that impacts are important on all of them.

So far we have treated impacts as single, random events. But clustering of the orbits of comets and asteroids is a demonstrated fact. Impact "storms" are a real possibility that we discuss in some detail in chapter 11.

11

COMET AND ASTEROID FAMILIES

There are several groups of two or more short-period comets whose orbits are so situated that it would be possible for all the comets of one group to be simultaneously very near one another and near Jupiter also. It is not unlikely that each of these groups has been produced by the disruption of a single comet at some past epoch. Several instances of the profound alteration of a comet's orbit by an encounter with Jupiter have actually occurred.

H. N. RUSSELL, R. S. DUGAN, AND J. Q. STEWART
Astronomy, 1926

1. The astronomical environment as we have described it leads one to expect as the prime process, not solitary random impacts but bombardment episodes, galactically controlled. These impose a temporal pattern on Earth processes.

2. Within an impact episode, events are dictated by a handful of rare giant comets. The disintegration of these bodies liberates copious amounts of dust into meteor streams, the dust spreading out to form dense, temporary zodiacal clouds which may be several hundred times as massive as the one existing at the present day. . . .

3. The Earth has run into swarms of bodies, each with the potential for numerous blasts and global conflagrations, as well as a number of larger Apollo asteroids.

VICTOR CLUBE AND BILL NAPIER
The Cosmic Winter, 1990

It has been known since 1865 that some comets cluster into families with very similar orbital properties. About two dozen such groups are now known. In particular, as Russell, Dugan, and Stewart pointed out in the passage quoted above, several of these groups have orbits that are so oriented that the comets in them could all simultaneously be close to one another and close to Jupiter. Jupiter, after the Sun, is the second most powerful gravitating body in our solar system. A comet that, for the sake of argument, grazes the top of Jupiter's atmosphere, is subject to gravitational tidal forces that are essentially equal to those that would be experienced by a comet that grazes the surface of the Sun. There is a major difference between these environments, however: a comet that comes close enough to the Sun to experience such strong tidal forces will also be subjected to extremely intense solar heating. Half the sky will be occupied by the fiercely hot surface of the Sun, at a temperature in excess of 6,000° C. The experience is likely to be fatal to a fragile ice ball. No detectable fragments may survive this harrowing experience. Yet comet families do exist!

One of these comet families is truly spectacular. It consists of the Great Comets of 1668, 1843, 1880, 1882, and 1887. At least two other known comets are probable members of this group. The orbits of these comets are clearly very similar, but the calculated orbital periods are so long (400 to 1,000 years) that there is no possibility that they are merely different perihelion passages of the same comet. The most spectacular member of this family, the Great Comet of 1882, split at or slightly after perihelion passage (0.002 AU from the solar surface!) into four smaller comets with orbital periods of 664, 769, 875, and 959 years.

Other comets have also been seen to break up. Biela's comet, with the unusually short orbital period of 6.7 years, split in half in 1846. Taylor's comet of 1916, with a 6.5 year orbital period, also split into two pieces. Comet expert Zdenek Sekanina of the Jet Propulsion Laboratory has collected all available observations of these events. Of the twenty-two documented cases of splitting of comets into two to five parts (producing fifty-five fragments), the highest parting velocities are only a few meters per second. It might seem that splitting should occur near the time of perihelion passage, when the nucleus is subjected to both strong tidal forces exerted by the Sun's gravity and strong surface heating by the Sun's radiant energy. However, the observed disruption events took place from well inside Earth's orbit out to Saturn's orbit! The mean distance from the Sun at the time of breakup is about 2 AU. Clearly comet nuclei are very weak.

The physical properties of comet nuclei are poorly understood, since only one (Halley) has even been examined by a flyby mission. That comet nuclei generally contain intermingled fine-grained dirt and ice in roughly equal masses is clear from a wide range of studies of the gases and dust expelled by active comets as they pass close to the Sun. However, the density of that material is poorly defined by these observations. Some nuclei may be fluffy assemblages of snowflakes and mineral grains with densities less than 0.1 grams per cubic centimeter (water's density is 1.0), containing at least 90 percent empty space. Others may be firmly compacted permafrost, with dirt and ice pressed free of void spaces, and a density of 1.5 to 2.0. The ease of breakup of comets suggests very low tensile strengths; however, a nucleus consisting of several dozen large blocks of dense, strong permafrost resting lightly upon each other in the comet's tiny gravity field, would also satisfy the requirement of easy breakup and give a low overall density.

From meteor studies we know that cometary meteors break up over a very wide range of dynamic pressures during atmospheric entry. Some centimeter- to meter-sized meteors are completely crushed by aerodynamic pressures as low as 0.01 atmospheres, whereas others may remain intact above 10 atmospheres dynamic pressure. The weakest, the type 3 meteors, are fairly described as cometary dustballs. Much stronger are the type 2 meteors, with the physical strength of carbonaceous chondrite meteorites, which they also resemble chemically. The strongest meteors, type 1, are asteroidal material similar to ordinary chondrites. It is quite certain that much of the ice-free material left behind by a dying comet is extremely weak and fluffy. It is easy to understand this dustball material as the result of gentle evaporation of ice from a porous ice-dust mixture.

However, the denser, stronger cometary stones must be derived from strongly compacted, nearly void-free solids. Some have even suggested that melting of ice and density-dependent separation of water and "mud" may have occurred inside comet nuclei, resulting in the formation by settling of cores of wet rocky materials generally similar to carbonaceous meteorites. Even if this happened, which is doubtful, it is no easy matter to establish a direct connection between carbonaceous solids produced in the cores of comet nuclei and carbonaceous meteorites in our museum collections of Earth. The carbonaceous meteorites are so weak that entry into Earth's atmosphere at cometary orbital speeds would crush and destroy them utterly.

Recalling our discussion of comet-asteroid relationships in chapter 6, there are two ways around this dilemma. The first is to have the carbonaceous meteorites derive directly from low-speed near-Earth asteroids that are extinct comet cores. Many such extinct comets must be present among the NEAs, and many of them must be in near-circular (low-eccentricity), low-inclination orbits with low atmospheric entry velocities. The course of orbital evolution from long-period to short-period orbits, and the slow evaporative loss of the last remaining ice, make bodies that are "asteroidal" in orbital and physical properties out of starting materials that are unabashedly cometary. Carbonaceous NEAs can have low enough entry speeds that they stand a reasonable chance of dropping survivable meteorites (ulitimately of cometary origin) through the atmosphere. The second route, actually somewhat simpler, is to have the carbonaceous meteorites made by similar process of melting and separation according to density, occurring not inside a long-period comet, but in ice-bearing asteroids in the outer belt. These asteroids may then be disturbed by gravitational perturbations by Jupiter into eccentric orbits, and then evolve into Earth-crossing orbits. In either case, the chemistry and physics of the products are very similar.

The logical implication of a cometary origin for roughly half of the NEAs is that physical studies of currently accessible asteroids may shed much light on the nature of cometary nuclei. First of all, it was discovered by Jack Drummond (who was then, in 1990, at the University of Arizona) that there are three rather well-defined orbital families among the Amor asteroids, reminiscent of the families of comets discussed above, and of the connections between comet orbits and meteor streams described in chapter 2. The first of these groups, imaginatively labeled Group 1, contains five NEAs, 1989 RC, 1972 RB, 1987 SF3, 1981 QA (alias asteroid #3102) and 1987 QB. Group 2 also contains five asteroids, 1983 RD (#3551), 1989 VB, 1980 PA (#3908), 2061 Anza, and 2202 Pele. Group 3 has four members, 1982 XB (#3757), 1980 AA, 1980 WF, and 1989 UP. The only tenable explanation of this statistically significant clustering of asteroid orbits is that each cluster is derived from the breakup of a single progenitor.

Breakup is easiest for already-fractured bodies with "rubble-pile" structure. Typically the biggest fragment produced by breakup is 10 to 50 percent of the total mass. To check breakup theory using observations of NEA masses requires that we know the density and volume of many NEAs. But it is difficult to estimate their volumes accurately because no spacecraft has yet visited an NEA. They have therefore never been photographed with sufficient resolution to show their sizes and shapes. Steve Ostro of the Jet Propulsion Laboratory has provided the very best images of NEAs by means of radar mapping of a few of these tiny objects that have strayed very close to Earth. In order to find an NEA on radar, it must have already been well enough observed so that its orbit is precisely known. (The beam transmitted by a planetary radar is so narrow that using it to search for new asteroids is utterly out of the question.) Secondly, the target asteroid must have a predicted close flyby of Earth in the near future.

Finally, the target must be both large enough and close enough so that a strong radar return can be attained, with enough signal strength so that, after splitting the signal up into a large number of delay and Doppler bins, there is still something to work with above the background noise level.

Over the past few years, Ostro has diligently pursued several of the most promising targets. The results are little short of astonishing. From the limited list of four well-observed NEAs, three are very elongated and complex in structure, more closely resembling pairs of rocks touching each other than single, roughly spherical bodies. Such objects, which touch each other and rotate so fast that the pieces are nearly in orbit around each other, are called contact binaries. Contact binaries of any size that rotate in less than one or two hours would simply fly apart. These NEAs appear to be nearly shattered, or perhaps shattered and nearly dispersed, bodies. It will be interesting to learn the compositions of the members of these binaries to see whether they are composed of the same material or are distinctly different in composition. If the members of a pair should be found to differ markedly in composition, this would be evidence for either melting and density-dependent separation of the parent asteroid (if they are compatible rock types), or of low-velocity collision between two bodies with independent origins.

On several occasions, observations of eclipses of stars by belt asteroids have found strange, brief occultations ("hidings"; nothing mystical here) of the star shortly before or after the main event. Diligent searches for mechanical or electronic glitches that might have produced false eclipses have failed to find any reasonable way to explain away the observation. It is therefore entirely possible that many asteroids have small satellites that orbit independently around them, not in contact with their surfaces.

In March of 1994, an image of the belt asteroid Ida was transmitted back to Earth by the *Galileo* spacecraft as it passed through the heart of the belt on its way to Jupiter. There, alongside Ida, was a tiny chunk of rock only a couple of kilometers in size. At last we have direct proof of the presence of satellites of asteroids.

Several major impact features on Earth, most notably the Clearwater Lakes in Canada, show two craters of the same age only a few crater diameters apart. Their separation is too large to be attributed to breakup inside Earth's atmosphere. Further, Jay Melosh and John Stansberry of the University of Arizona have noted that close doublet craters on Earth are much more common than mere chance would allow. The easiest explanation of crater duplicity is that the impacting body was an asteroid with a close satellite. There is an interesting process at work here: contact binaries would generally not produce a visibly double crater, because crater diameters are typically several times the diameter of the impactor: the subtle details of structure get lost in the scope and violence of the explosion. Contact binaries would produce a single, round crater.

We should also recall that asteroids have very weak gravity fields. For a satellite to persist in orbit for a prolonged period of time (and thus have some chance of being observed), it must not stray too far from the asteroid around which it revolves, or gravitational tidal forces exerted during close flybys of planets would strip the satellites off and send them into independent orbits around the Sun. Note that approaches to within ten planetary radii are one hundred times more common than actual impacts.

One possible case of paired impacts has been widely discussed in connection with the Cretaceous extinction event. Until late 1993, there were two known craters with dates compatible with the K/T boundary impact, leading to speculation that the K/T "event" may have been an impact shower or a binary asteroid. Several paleontologists have emphasized that the extinctions *near* the K/T boundary may not all be *at* the boundary. The extinctions, they claim, appear from the stratigraphic evidence to be diachronic (spread out over time). This, plus the possible contemporaneity of Manson and Chicxulub, naturally spurred interest in the possibility that a family of impacts of different sizes occurred over a million years or so near the time of the Chicxulub impact, the one that left the boundary clay layer. But accurate dating of the Manson structure in Iowa has now shown that it is clearly several million years too old to be correlated with the boundary extinctions and the Chicxulub crater. The more accurate dating of Manson removes the basis for their conjectured connection.

Given all the evidence that comets form orbital families, that meteor streams often correlate with known comets, that NEAs have orbital families, that some meteor streams are related to NEAs, that contact and noncontact duplicity are apparently common among asteroids, and that about half the NEA population is of cometary origin, the discovery of multiple impacts clustered in time would not have come as a great surprise. Statistical studies of both the age distribution of impact craters and the times of biological extinctions have suggested that both show clustering. According to this claim, the clusters are not themselves randomly distributed in time, but fall into a regular pattern with a roughly 26-million-year periodicity. Further, some of these studies claim that the pattern of extinctions and the pattern of cratering are in step with each other. Clearly, then, a causal relationship would be plausible. When we find a correlation of dead bodies with smoking guns, it is difficult to avoid the suspicion that there is some connection. Since extinctions cannot cause impacts, the most economical explanation seems to be that the impacts caused the extinctions!

As an attempt to explain the claimed periodicity of impacts, D. P. Whitmire and A. A. Jackson suggested that there is a faint star in highly eccentric orbit around the Sun with a period of 26 million years. The alleged solar companion star was named Nemesis by its advocates. The orbit of such a companion star would not be stable over time periods of 100 million years because gravitational disturbances by passing stars and giant molecular clouds would quickly distort the orbit to a very different period, or cause Nemesis to escape completely from the Sun's gravity. But, on a more fundamental level, many scientists have reservations about the significance of these alleged periodicities and correlations, based upon the small number of events used in reaching the conclusions. My personal reaction is to return a verdict of "not proved," which is allowed in some legal systems, and to avoid a categorical "guilty" or "not guilty" verdict.

Whether or not one accepts the notion of regular periodic impact storms, the idea of episodic clusters of impacts remains highly plausible. A scenario built by Victor Clube and Bill Napier of Oxford University favors the idea of periodic extinctions, but rejects the Nemesis hypothesis as untenable. They propose in their book, *The Cosmic Serpent*, that periodicity is imposed on the impact rate by the periodic oscillation of the Sun through the plane of the Milky Way galaxy as it pursues its orbit about the galactic center. Each passage of the solar system through the galactic plane is likely to be disruptive of the Oort cloud of comets around the Sun. Many large comets would be perturbed into orbits that penetrate the inner solar system. Each such large comet, once it has encountered Jupiter and been perturbed into a short-period orbit, may then shatter into a hail of fragments with dynamically similar orbits, many of which may present a hazard to Earth.

As an example of such an event in the relatively recent past, Clube and Napier cite Encke's comet and the elaborate system of meteor streams, fireballs, and other debris associated with it. Many of the

fireballs from this family have been in short-period, NEA-like orbits long enough for extensive precession of their long axes around the Sun. Since the orbits are similar, not identical, the various individuals precess at different rates. As time passes, the points of intersection of these orbits with Earth's orbit drift around the Sun at different rates. At present the main body of Enke-related meteors and fireballs, the Taurid stream, is oriented so that it intersects Earth from mid-October to mid-December. Three August and September meteor streams also are clearly related. The other crossing point of Earth's orbit results in fireball activity from March through May. This entire set of meteor streams and fireball families is probably derived from several bodies in Encke-like orbits, one of which (Enke itself) is still active today. These bodies are collectively called the Taurid complex.

We can only speculate regarding the ultimate parent of these orbitally related comets. The simplest explanation is that a dozen or so kilometer-sized comets were derived from a single larger comet nucleus, which broke up in the same manner as the other comets we have observed to split.

Several NEAs, including 1990 SM, 2212 Hephaistos, 4197 1982 TA, and 1993 KA2, have orbits that are similar to the orbit of Enke's comet. It is possible that they are dormant fragments of comet cores from earlier breakup episodes of Enke or Enke's parent. These orbits have precessed apart, attesting to the passage of many millennia since they all traveled together as a compact swarm. Shortly after the fragmentation of such a comet or asteroid, the fragments would all be following closely similar, closely aligned orbits that would cross Earth's orbit on one or two precisely defined dates each year. Thus it would be possible for an observer with years' worth of data to predict the likely recurrence of spectacular meteor showers, or even meteorite falls, on the same date each year. It may be that this is what Anaxagoras really did.

Napier and Clube argue forcefully that the accepted cratering rate in the solar system is actually a long-term average that does not have enough time resolution to discern the many brief, violent episodes that dominate the cratering history. The actual rate at any time is usually much less than the average, but the relatively brief, intense bombardment episodes that follow the breakup of a giant comet may have an impact rate one hundred or more times the average. They further suggest that we are presently living in a time of enhanced cratering activity.

Their suggestion requires the presence of giant comets in the outer solar system. One might rightly wonder whether any such bodies have been found. In 1977 a monster iceball named Chiron, which impertinently crosses the orbits of the giant planets, was discovered. Such behavior cannot for long go unpunished by nature (or, for that matter, unpublished by *Nature*). Computer predictions of Chiron's orbit reveal, not surprisingly, that very strong gravitational perturbations of Chiron's orbit by these massive planets are common. As a result, Chiron's orbit is *chaotic* in the long term. The simulations done to date suggest that Chiron, a 200-km body with ten thousand times the mass of the Cretaceous impactor, will wander into an Earth-crossing orbit on a time scale of one hundred thousand years. The discovery by Spacewatch of two other such bodies, the 200-kilometer Pholus (1992 AD) and 80-kilometer 1993 HA2, in January and April of 1992, has shown that other Chiron-type bodies (now collectively called Centaurs) are present among the Jovian planets. Unfortunately, there is as yet no basis for estimating their total number or total mass.

But 1993 brought an even more striking discovery with strong impact on this discussion: This entry was made by Brian G. Marsden in *IAU Circular 5800* on May 22, 1993:

> PERIODIC COMET SHOEMAKER-LEVY 9
>
> *Computations by both Nakano and the undersigned were beginning to indicate that the presumed close encounter with Jupiter occurred during the first half of July 1992, and that there will be another close encounter with Jupiter around the end of July 1994. . . . This computation indicates that the comet's minimum distance from the center of Jupiter was 0.0008 AU (i.e., within the Roche limit) on 1992 July 8.8 UT and that Delta J will be only 0.0003 AU (Jupiter's radius being 0.0005 AU) on 1994 July 25.4.*

On March 24, 1993, Carolyn and Gene Shoemaker and David Levy, a team of experienced discoverers of comets and asteroids, found a very peculiar comet-like smear of light on one of their photographic plates. Careful inspection of the image showed what looked like a string of pearls embedded in a bright haze. The discovery was immediately reported to the International Astronomical Union's clearinghouse for hot new discoveries, Brian Marsden's Central Bureau for Astronomical Telegrams at the Smithsonian Astrophysical Observatory in Cambridge, Massachusetts. Alerted by Marsden's telegrams (actually, nowadays it is electronic mail that does most of the work) a number of astronomers around the world hastened to follow up on this strange body, called a comet for lack of a better name. Images taken by several observers at different observatories revealed as many as twenty-one "pearls" in a straight line, with a highly asymmetric haze of light around them. Searches for the emission lines and bands of many common cometary molecules failed: there was no gas present. The haze must therefore be reflecting dust, not glowing gas. The "comet" thus more closely resembles a disrupted asteroid or satellite than a true comet.

As observations continued over the course of several weeks, it became possible first to calculate a crude orbit for the "comet" about the Sun. This calculation showed an orbital period very similar to Jupiter's. It also revealed, as in the quotation from Marsden given above, that an extremely close approach to Jupiter had occurred the previous July. At that time, the comet passed so close to Jupiter that it cut inside the critical distance at which the gravity of Jupiter would have as strong an effect on a piece of the comet's surface as the body's own gravity. This point, called the Roche limit after the French astronomer who discovered the principle, marks the point at which an approaching body with little cohesive strength, held together only by its own gravity, will be pulled apart by Jupiter's gravity. The "string of pearls" consists of a chain of fragments of a multikilometer cometary or asteroidal body that broke up in 1992. But Marsden's statement went beyond this already interesting fact: he reported that the body was in an extremely elongated orbit around Jupiter, not the Sun, and was due to return to close proximity with the giant planet in July of 1994. The detailed numbers describing the coming encounter indicated that its closest approach to the center of Jupiter would be less than the radius of Jupiter. In effect, he announced to the world the impending collision of the bizarre "comet" Shoemaker-Levy 9 with Jupiter.

This was not the first observed example of a comet being captured by Jupiter: the short-period comet Gehrels 3 had been temporarily captured and then lost after a brief interval. But Shoemaker-Levy 9 was the first observed example of capture leading to impact.

Since the spring of 1993 a great body of observations and theory have sprung up around this unprecedented event. Not since Gervase's report of an impact on the Moon in June of 1178 has anyone on Earth witnessed a comet or asteroid impact on another planet. It promised, in a very real sense, a test case for our theories of the behavior of impactors. And that was the good news.

The bad news was that the impacts of the individual fragments would occur on the dark side of Jupiter, out of direct view from Earth. Instruments aboard both the retreating *Voyager* and the approaching *Galileo* probe could view the impact site, but only from a great distance. There seemed to be a good chance that the flash of the explosion would be so bright that observers on Earth would see Jupiter's Galilean satellites brighten briefly as each of the larger fragments augered in. If we were lucky, the speculation ran, we might be able to see the fireball of an impact rise above the horizon. With an impact speed over fifty kilometers per second, the energy content of the largest fragments must be tremendous, similar to a long-period comet impact on Earth. Some astronomers estimated that, if the impact were on the dayside, it would be an easy naked-eye event for observers on Earth. In any case, for the few days that it would take for the entire string of pearls to fall into Jupiter, millions of people would be watching to see a true spectacle of nature.

The sizes of the individual fragments were still poorly known, but the largest was estimated to be anywhere from half a kilometer to perhaps three or four kilometers in diameter. The kinetic energy of a body of three kilometers diameter at the moment of impact on Jupiter would be about 10 million megatons, compared to five megatons for the explosion of Mount St. Helens. Much greater uncertainty surrounded the question of how deeply Shoemaker-Levy 9 would penetrate Jupiter's atmosphere before exploding. Proponents of the most extreme "weak-comet" ideas expected complete disintegration at very high altitudes, with negligible effect on the cloud layers and atmospheric circulation. The "hard-rock" extremists thought that the larger bodies might penetrate all the way to depths where the atmospheric pressure is several times as high as on Earth, causing explosions that would disrupt the cloud layers over a vast area and disturb the large-scale circulation of the atmosphere. When the day of reckoning finally arrived, the truth was found to lie well in the middle ground between these extremes.

The events of late July were certainly spectacular enough to satisfy even the most jaded critic. The explosions of the larger bodies on the back side of Jupiter launched huge fireballs thousands of kilometers above the rim of the planet, into plain view from Earth. As the string of impactors dove in like runaway cars from an uncoupled train, one blast after another could be clearly seen by observers on Earth. The larger blasts left immense black smudges of dust on Jupiter's cloud-covered face. Several of the larger impact scars are much bigger than the entire surface area of Earth, even though dispersing these dust clouds on Jupiter requires fighting a gravitational acceleration some two and a half times as large as on Earth.

The best present estimates of the total explosive yield are tens of millions of megatons. It is likely that, once the data from the *Galileo* spacecraft has trickled back to Earth over the coming months, we will have a much more precise picture of the nature of the explosions, their total explosive power, and their depth of penetration into Jupiter's deep atmosphere.

Since the discovery of Shoemaker-Levy 9, calculations by Gene Shoemaker and others have suggested that a cometary breakup event should occur near Jupiter at the rate of about one per century. This calculation suggests that freshly fragmented cometary and asteroidal bodies exiting the Jupiter system like strings of pearls may occasionally collide with one of the Galilean satellites. Jay Melosh has reexamined the *Voyager* photographs of the Galilean satellites and identified at least a dozen examples of strange chains of impact craters on Ganymede (the biggest target) and Callisto. Should such a string of fragments be directed toward Earth, it is very unlikely that more than one would strike on a given passage through the inner solar system. But the others would present a continuing hazard to Earth for thousands of years to come.

Which brings us to the best news of all: the impact was on Jupiter, not Earth . . . *this* time. We could reasonably expect a decent interval of peace in which we might think about these catastrophic events. But this was not to be: within weeks after the SL-9 impacts, the new comet Machholz 2 was observed to split into three large pieces. These three comets are all on trajectories that take them into the inner solar system, as close as 0.3 AU from Earth. In the long term, as these orbits evolve, they will constitute an impact hazard to Earth.

12

CRATERS IN THE OCEAN DON'T LAST

The Earth's axis and rotational motion changed; the oceans abandoned their former beds, to rush toward the new equator: the majority of men and animals were overwhelmed by this universal deluge, or destroyed by the violent shock; entire species were annihilated; every monument of human industry overthrown: such are the disasters which might result from collision with a planet.

PIERRE-SIMON LAPLACE, 1793

Waves caused by ocean impacts may be the most serious problem produced by impacting asteroids short of the massive killers such as the Cretaceous/Tertiary impactor. . . . An asteroid with a radius of 200 meters that drops anywhere in the mid Atlantic will produce deep-water waves that are at least 5 meters high when they reach both the European and North American coasts. When it encounters land, this wave steepens into a tsunami over 200 meters in height that hits the coast with a pulse duration of at least 2 minutes. . . .

Because a disproportionate fraction of human resources are close to the coasts, tsunamis are probably the most deadly manifestations of asteroid impacts apart from the very large . . . superkillers.

JACK HILLS AND PATRICK GODA
Astronomical Journal, March 1993

Everyone has heard of *tsunamis*, but not necessarily by that name. They are the giant "tidal waves" that are the scourge of the Pacific basin, a threat to life and property—and they have absolutely nothing to do with tides. Tsunamis are caused by violent oceanic events such as earthquakes, undersea landslides, volcanic explosions, and impacts. Some of these giant waves reach extraordinary heights. There are tidal wave deposits over three hundred meters (one hundred stories) above sea level, far up the slopes of the island of Lanai in Hawaii. Smaller tidal waves frequently wash low-lying coastal areas throughout the Pacific. We have seen television reports on evacuations in California, Alaska, Hawaii, and Japan in the face of approaching tidal waves less than a meter high. Many of us have read James Michener's fic-

. . . .

tional account of an earthquake-caused tsunami wave sweeping the shore in *Hawaii*. The futurists among us have also read Larry Niven and Jerry Pournelle's account of a giant tsunami, generated by an impact of a comet nucleus in the Pacific Ocean, sweeping the Los Angeles basin. Their story, complete with maniacal surfers who succeed in catching the "ultimate wave" and riding it far above Santa Monica Boulevard, is no mere Alvarez exploitation. No, indeed: *Lucifer's Hammer* was written in 1977, three years *before* the Alvarez team discovered the K/T boundary iridium layer . . . and three years *after* Arthur C. Clarke demolished Padua, Venice, and Verona in *Rendezvous With Rama*. After the iridium story emerged, writing cosmic-impact novels became a thriving cottage industry. Nonetheless, no writer of fiction has visualized the effects of giant ocean impacts as well as Niven and Pournelle.

Three out of every four impacts on Earth hit water. As on land, ocean impact explosions excavate huge craters. The fireball of an impact in deep ocean water can open a transient cavity tens of kilometers in diameter, reaching down all the way to the ocean floor, five kilometers deep. The water displaced from the explosion cavity is partly ejected in a broad, open cone at many times the speed of sound. The seabed is cracked by the blast wave, melted and scoured by the one-hundred-thousand-degree fireball. Hundreds of cubic kilometers of water are vaporized, blasting an immense column of steam back out to space, filling the spray cone with steam at temperatures of many thousands of degrees and pressures of thousands of atmospheres. When the surface of the fireball coasts to a stop in the water, the ocean surface collapses back into the cavity as a wave over five kilometers high. The wave accelerates into the nearly circular cavity from all sides, converging on the center of the crater. As the wave crest approaches the center of the crater, fast-moving waves converging from all directions pile into each other, rushing headlong into a monstrous surge that shoots up a towering pillar of water higher than the highest mountains on Earth. The sea sloshes back and forth in the blast region, pumping the surrounding ocean and generating circular wavefronts which, like the ripples from a pebble tossed into a puddle, spread out in all directions.

Because of the enormous energy content of tsunami waves, their behavior is a little different from what we see in puddles. The speed of a typical tsunami is about 750 kilometers per hour (450 mph), nearly two-thirds of the speed of sound in air. If you see one coming, it's already too late to get out of the way. The famous explosion of the Indonesian volcano Krakatau, on August 26, 1883, raised a 40-meter (135-foot) tidal wave that ran 1 to 10 kilometers inland along 500 kilometers of coastline, killing over forty thousand people. The one-hundred-megaton blast of Krakatau's explosion was clearly heard in Mauritius and Madagascar, Australia and New Zealand, out to a distance of well over 5,000 kilometers. Nine hours after the blast, the tidal wave sank hundreds of fishing boats in the harbor of Calcutta, India. The wave was clearly seen and felt in the English Channel thirty-two hours later! The sound of the blast reverberated around the world for nine days. A global pall of dust raised by the eruption reddened sunsets everywhere on Earth for several months.

Indonesia and the Philippines remain very hazardous places: as recently as August 23, 1976, a submarine earthquake in the Celebes Sea hurled a tsunami ashore on Mindanao, killing eight thousand.

Even the devastating forty-meter wave from the explosion of Krakatau is no record: in 1737 a sixty-four-meter (twenty-one-story) tsunami ran ashore on desolate Cape Lopatka on the Kamchatka Peninsula. And then there is that tsunami deposit found three hundred meters above sea level on Lanai!

The ocean surface transmits such giant waves because the behavior of water-surface waves is very different from the behavior of blast waves in air. Aerial explosions spread their energy out over the surface of an expanding sphere the surface area of which increases with the square of the radius. The speed of expansion, after an initial spike, is roughly constant. The blast wave in air therefore has an intensity (force per unit surface area) that drops off with the square of the time since the explosion: an explosion that delivers a blast wave pressure pulse of one hundred atmospheres after one second will pack a punch of only one atmosphere after ten seconds, when the radius of the blast wave is ten times larger. But waves on the surface of the ocean spread out in two dimensions, not three. The wave power is not spread around the surface of a sphere, but around the periphery of a circle. The strength of the wave therefore drops off inversely with the distance it has traveled, not as the inverse square of the distance. Further, deep-water surface waves propagate with almost no frictional loss of energy. Because of these fundamental properties of ocean surface waves, they are capable of delivering their punch over enormous distances.

In the Pacific basin, the long-distance striking power of tsunami waves can be seen to full advantage. Consider the major earthquake that occurred in Chile in May of 1960. The earthquake triggered submarine landslides that dislodged hundreds of cubic kilometers of sediment on the continental slope. The water displaced by these landslides drove a tsunami wave that expanded out over the entire Pacific. After traveling a quarter of the way around the world, at the time that the wave had the greatest possible radius (equal to the radius of Earth), the wave struck Hawaii from the southeast. As the wave, traveling 750 kilometers per hour, encountered shallower water near the coastline of Hawaii, frictional interaction of the wave with the bottom slowed the leading edge of the wave and held back the bottom of the surge. The wave rapidly grew from a height of tens of centimeters to an average of about 10 meters. The three-story-tall wave smashed into the southeast corner of the island of Hawaii, ravaging Hilo harbor, and killing sixty-one people. Continuing on past Hawaii, the wave had passed more than a quarter of Earth's circumference. From there on, the wave packed its energy into a wavefront that was converging toward Chile's antipodal point and growing in strength.

Twenty-three hours after the Chilean earthquake, and over seventeen thousand kilometers from its source, the tsunami arrived in Japan. The open-ocean wave was only some twenty centimeters high, but the long, shallow-sloping wave began to pile up as its base encountered the shallows off the Japanese coast and slowed down. The wave ran ashore with a height that varied from one to five meters, depending on the local coastal geometry. The inundation killed 114 people in Japan, and left another 90 missing.

Other major tsunamis on the exposed west coast of South America, driven by deep earthquakes caused by the breakup of the oceanic crust as it is forced under the edge of the continent by continental drift, were recorded in 1562, 1570, 1575, 1604, 1657, 1730, 1751, 1819, 1835, 1868, 1906, and 1922. I strongly recommend the gripping eyewitness account of the 1868 Chilean tsunami reprinted in the 1988 book *The Great Waves*, by Douglas Myles.

Japan has been wracked by numerous deadly tsunamis. The combination of violent earthquakes, submarine landslides, and volcanism renders all of Japan unstable, and the high population density and ancient culture of Japan have favored the keeping of records of their devastation. Over 1,000 people were killed by a tsunami in the year 869 on the Sanriku coast of northeast Honshū. Another tsunami in 1293 killed about 30,000. In 1611 a twenty-meter (seven-story) wave came ashore at Yamada Bay, killing 5,000. Another great wave in the same area in 1703 is said to have killed over 100,000. Thousands more were killed by tsunamis in 1717 and again in 1854. In 1896, a tsunami on the Sanriku coast left 27,122 dead. That wave peaked at nine stories above sea level at Yoshihama. And on March 3, 1933, an undersea earthquake drove a twenty-four-meter wave ashore in the same area, killing 3,000 more.

Hawaii, located in the mid-Pacific, receives moderate-sized tsunami waves from the "ring of fire" of the Pacific Rim. In addition, submarine landslides on the slopes of Hawaii's own gigantic volcanoes can generate extremely intense local waves. In the 165 years from 1813 to 1978 Hawaii recorded no fewer than ninety-five tsunamis. One of the best reported disturbances, the Aleutian earthquake of April 1, 1946, sent a towering wave into the harbor of Hilo, killing 159.

But tsunamis are not limited to the Pacific basin. On June 7, 1692, a powerful tsunami wave struck Jamaica, killing two thousand. A great earthquake on November 1, 1755, caused by the collision of the African crustal plate with Iberia, sent a succession of huge tsunami waves ashore along the coast of Portugal and southeastern Spain. Lisbon was leveled by the quake, then burned. The fires combined into a raging firestorm, driving the survivors toward the seawall, where they were struck by the waves. Some seventy thousand were killed in Lisbon alone. Perhaps ten thousand more were killed by tsunami waves running up on the coast of Morocco. A wave from this earthquake reached a height of seven meters across the Atlantic on the island of Saba in the Leeward Islands, more than six thousand kilometers from its source.

Several other deadly tsunamis have been reported in the last few years. On September 1, 1992, an earthquake off the coast of Nicaragua sent a 1.5-meter wave ashore at the port of San Juan del Sur, killing 170. Well over 10,000 Nicaraguans were left homeless by this very modest wave. In December 1992, an earthquake off Flores Island, Indonesia, sent a tsunami ashore along a populated stretch of coast, killing over 1,000 people and demolishing several villages. An earthquake in the Sea of Japan on July 12, 1993, sent tsunami waves with heights up to 30 meters (97 feet) ashore, demolishing the town of Aonae and killing 120 people.

The destruction wrought by tidal waves can be immense because of several factors. First, as the example of the Chilean earthquake tsunamis of 1960 illustrates, tsunamis may have global range. Second, the increase in height of tsunami waves as they run up onto continental shelves, like the waves that struck Hawaii and Japan in the Chilean example, is often a factor of 30 or 40, and sometimes even as high as a factor of 120. Thus, very long-wavelength ocean waves with heights of only fifty centimeters (twenty inches), which may not even be noticed by ships at sea, will strike a coastline with a typical height of fifteen to twenty meters, about that of a five- or six-story building. And third, in the last few centuries there has been a major shift of population to seaside cities.

The presence of major population centers very close to mean sea level has been much discussed in connection with global climate change. Global warming causes a rapid retreat of the great continental ice sheets in Antarctica and Greenland, causing the oceans to rise possibly tens of meters. (Melting of pack ice in the Arctic Ocean would be visually spectacular, but would result in no change of sea level: the ice is already floating in the ocean! Likewise, alpine-type glaciers in high mountain ranges all over the world would rapidly melt, but the total volume of ice in them is of little importance compared with the Antarctic and Greenland ice caps.)

The populated areas most immediately at risk from inundation by oceanic flooding or tidal waves of course include the Netherlands, the coast of Bangladesh, and the Atlantic coastal plains of North and South America. But consider a few examples: the altitude of much of Rome, Philadelphia, and Montreal is about thirty meters; Berlin is at thirty-five meters; San Francisco, New York, and Halifax are near twenty meters; Tokyo, Sydney, Hilo, and Rio de Janeiro are at about eight to ten meters; Boston, Honolulu, Tampa/St. Petersburg, Baltimore, San Diego, Savannah, Jacksonville, and Washington, D.C., are mostly close to seven meters above sea level. Miami, New Orleans, Seattle, Alexandria, Amsterdam, and Calcutta are about three meters above sea level. Even modest elevations of sea level therefore can threaten many major population centers.

Many large cities are built on slopes at seaside with varying elevations. They are no less at risk, and include Hong Kong, Kyoto, Osaka, Lisbon, Bombay, Los Angeles, Mobile, Galveston, Madras, Shanghai, Vladivostok, St. Petersburg (Russia), Colombo, Singapore, Auckland, Santiago, Havana, Vancouver, Quebec, Anchorage, Edinburgh, Copenhagen, Athens (Piraeus), Istanbul, Manila, Dublin, Haifa, and Bangkok, to list a few. Many cities are high above sea level but located on the shores of lakes or seas that present excellent targets for impacts: Chicago, Detroit, Erie, Cleveland, Milwaukee, Toronto, Odessa, Kampala, Thunder Bay, Duluth, Baku, et cetera. Although the bodies of water upon which these cities are built cover only a tiny fraction of Earth's surface area, they present a vastly larger target than the cities themselves. Thus a large impact anywhere on these bodies of water would be as devastating to neighboring cities as if the impactor were to hit the city directly. In addition, the capitals of many nations lie along the ocean: in West Africa, a three-thousand-kilometer stretch of coastline contains the capitals of Senegal, Gambia, Guinea-Bissau, Guinea, Sierra Leone, Liberia, Ivory Coast, Ghana, Togo, Dahomey, and Nigeria.

But is it likely that an impact-driven tidal wave with a runup height of ten or twenty meters will strike within foreseeable human history? How rare are such events?

There are two ways to seek answers to this question. The first is to look for geological records of ancient tsunamis in coastal areas throughout the world. The second is to use all the information we now have about the frequency of comet and asteroid impacts on Earth, the size distribution of the impactors, and their physical properties (that is, their breakup behavior in the atmosphere) to calculate the magnitude of the threat.

The geological record, unfortunately, is not unambiguous. A major tsunami will deposit broken trees near the high-water mark and move prodigious amounts of sediment. But, in most places on Earth, the organic debris will rot away in a rather short period of time, and normal erosion will remove some of the evidence (seashore litter, such as shells, carried to locations well above sea level) and bury most of the rest. Worse than that, there is nothing about such a deposit that tells us definitively whether it was made by an earthquake tsunami, a landslide tsunami, abnormal storm waves, or an impact tsunami. Subtle aspects of the sediment from the immediate locale of an impact may carry valuable clues, such as shocked quartz grains, tiny spherules of impactite glass blasted off the ocean floor, and debris from the impactor itself. Tsunami deposits remote from the impact site will carry no record whatsoever of nonterrestrial material. To date, not a single suspected tsunami deposit has been searched for such evidence.

The best impact target on Earth is the Pacific Ocean. Unfortunately, the Pacific basin is also surrounded by the volcanic "ring of fire," which follows the circle of subduction zones where oceanic crust is being forced down into the mantle under the continental margins by continental drift. The ring of fire contains a very large fraction of the earthquake activity of the entire planet. The 1960 Chilean earthquake discussed above is one example of many Pacific earthquakes that have generated tsunamis. Thus the evidence for tsunamis in the Pacific does not generally tell us the cause of the tidal waves. Since large earthquakes and submarine landslides up to at least magnitude eight are much more common than impacts of the same energy, impact-driven tsunamis cannot be very important in this size range. But there are natural limits on how powerful an earthquake can be, and hence on how powerful a tsunami an earthquake can generate. Most seismologists expect that there is a natural limit to the size of terrestrial earthquakes set by the depth and length of the biggest fault systems and the energy stored in them. Seismologist Carl Kisslinger of the University of Colorado has suggested that the natural limit on the size of earthquakes is around magnitude 9. The highest defensible estimates are several times higher in energy, perhaps magnitude 9.3 to 9.5. There seems no basis for any natural earthquake approaching magnitude 10. But magnitude 9 shocks from impactors occur every eighty thousand years, and magnitude 10 blasts every five hundred thousand years, and magnitude 11 shocks every 3 million years . . . and so on. Even within the short history of Homo sapiens, the most violent events on Earth have been extraterrestrial impacts. Over time scales of one hundred thousand years and longer, the greatest tsunami waves produced on Earth must be from cosmic impacts.

Consider a 1,000-megaton impact at a distance of 1,000 kilometers from the nearest coastline. An impact event of this size occurs every ten thousand years on the average somewhere on Earth. The open-ocean wave at a distance of 1,000 kilometers from the impact will have a height of about 4.5 meters before beginning its runup onto the shore. For a typical runup of a factor of thirty-five in height, the wave will run ashore with a height of 157 meters (520 feet). In any populated area, such a wave would cause total destruction. In flat countryside such as the Low Countries, the Yellow River valley, or coastal plains in the Americas, the wave would run far inland, destroying all human works and flooding rich cropland with saltwater. Waves with a tenth of this energy (waves from 100-megaton explosions in the ocean) are about six times as common, with average times between events of sixteen hundred years. These smaller explosions will raise open-ocean waves that are about a third of the height of those from the 1000-megaton explosion at the same distance, with average runup heights of 52 meters (170 feet; 17 stories) on shorelines 1,000 kilometers away.

Explosions in the one- to ten-megaton range are of course much more common; however, all but the very strongest bodies in this size range are vulnerable, like the Tunguska body, to fragmentation in the atmosphere. The principal hazards from impactors with explosive yields less than ten megatons are therefore blast-wave damage and firestorm ignition. Atmospheric explosions will at best generate rather feeble tidal waves.

Because of the nature of oceanic impacts, close observers of large impacts cannot survive. Over historical time, distant observers may have seen, and even survived, tsunami waves, but could have known nothing of the cause of the disturbance. Smaller impacts, such as meteorite falls into water, can be (and have been) both witnessed and survived.

A tenth-century manuscript, the *Chronicon Benedicti: Monachi Sancti Andreae in Monte Soracte*, reports such a fall: "In the year 921, in the time when the lord John X was pope, in the seventh year of his pontificate, a sign was seen. At that time, many stones were seen to fall from heaven near the city of Rome . . . in the village that is called Narnia. . . . The largest among all of these many stones fell into the river Narnus, and can still be seen to this day protruding to a height of about one cubit above the level of the river water."

A large meteorite fell on the banks of a river in China in 1369, creating havoc described in more detail in the next chapter.

In the English Channel off Dover, England, on December 17, 1852, a spectacular fireball was seen diving steeply toward the water. A loud sonic boom was heard by observers on shore. The fireball plowed into the water about a half mile offshore, raising a high plume of spray that was ripped to shreds by gale-force winds.

Near midnight of February 24, 1885, at a latitude of 37° N and a longitude of 170° 15'E in the North Pacific, the crew of the ship *Innerwich*, en route from Japan to Vancouver, saw the sky turn fiery red: "A large mass of fire appeared over the vessel, completely blinding the spectators; and, as it fell into the sea some 50 yards to leeward, it caused a hissing sound, which was heard above the blast, and made the vessel quiver from stem to stern. Hardly had this disappeared, when a lowering mass of white foam was seen rapidly approaching the vessel. The noise from the advancing volume of water is described as deafening. The barque was struck flat aback; but, before there was time to touch a brace, the sails had filled again, and the roaring white sea had passed ahead."

On August 20, 1907, the steamship *Cambrian* arrived in Boston from England with an extraordinary tale to tell. When the ship was several hundred miles south of Cape Race, Newfoundland, steaming along under a clear sky, a brilliant fireball appeared near the northeastern horizon and "rushed across the sky like a rocket. The next moment it passed over the topmast of the liner with a tremendous roar and plowed up the sea about fifty yards from the boat. The upheaval of the water was terrific, but the ship was not damaged." As far as I can see, fifty yards was all that separated this story from being a report of the mysterious loss of a ship and its crew.

On September 4, 1910, a meteorite plunged into a lake at St. Paul, Oregon, hurling a tremendous column of water high into the air. On August 29, 1913, in Fall River, Massachusetts, another meteorite fell in water with great commotion, "producing an explosion that sounded like the discharge of a twelve-inch gun" that broke windows and was heard for twenty miles. On November 27, 1919, a meteorite fell into Lake Michigan near the Michigan shore. "Residents of Battle Creek, Kalamazoo, South Bend, Grand Haven, and other Western Michigan cities fled from their homes in panic, fearing an earthquake. Houses were shaken, the country was illuminated as by a bright sun's rays, so all-enveloping it was impossible to tell from which direction the flare came, the earth trembled for half a moment and then came a deep prolonged rumbling as of a terrific explosion."

On April 23, 1922, the beach patrol in Toms River, New Jersey, reported seeing a brilliant fireball fall into Barnegat Bay. Huge rolling waves came ashore several minutes later. On the night of June 11, 1929, a fireball fell into Lake Superior. "The meteor illuminated the island brilliantly. . . . Quite a sea arose after the meteor struck the water." Then, on September 13, 1930, a fireball plunged into the sea near Eureka, California, "barely missing the tug Humboldt, which was towing the Norwegian motorship Childar to open sea."

"*World's End Feared.*" On October 19, 1936 a number of spectacular daytime meteors were reported falling into the sea at different locations in eastern Newfoundland at about 2:00 P.M. There were several reports of heavy explosions and impacts in bays many miles apart. "So brilliant was the illumination that, even in midday, people rushed to the windows in alarm." One was reported to have fallen in Fortune Bay, where another bright fireball had fallen the previous month.

On April 18, 1979, a meteorite was reported to fall into the waters of Barnegat Bay near Lanoka Harbor, New Jersey, only five miles from the site of the reported Toms River fall of 1922.

On February 15, 1988, at Rivers Inlet, British Columbia, one observer heard a loud aerial explosion followed by an impact in the water about three kilometers away and the appearance of a great plume of rising steam and spray.

I N the same way that impacts in the ocean generate a powerful conical blast and leave a deep transient crater in the water, so also impacts upon planets with atmospheres excavate a deep conical hole. Explosions of one hundred megatons or more inflate fireballs so huge that they break out of the atmosphere and open such holes. Sufficiently large explosions, in the million-megaton (one-thousand-gigatons; one-teraton) range, can accelerate some of Earth's atmosphere to escape velocity. Loss of atmospheric gases and impactor materials to space becomes even more important for larger explosions. In the 100-million-megaton class, the impact essentially scours off the entire mass of atmosphere above a plane that is tangent to the surface at the point of impact. Earth's powerful gravity prevents any but the most energetic impacts from removing more mass than it brings in.

The case is very different on Mars. The escape velocity of Mars, five kilometers per second, is less than half of Earth's. Further, the atmosphere is so tenuous that it offers little resistance to the expansion of the impact fireball. Even modest-sized impactors can blast atmospheric gases off of Mars at speeds above escape velocity. This process is called atmospheric erosion or explosive blowoff. When the conical jet of extremely hot gases accelerates to escape velocity, it readily sweeps away the atmosphere above the horizon plane. The gases, moving at speeds above escape velocity, can erode the planetary surface around the impact point, pluck rocks from the surface, and accelerate them relatively gently to escape. They then enter independent orbits around the Sun, eventually to become meteorites. Indeed, three separate classes of meteorites are known that contain Martian atmospheric gases. They are the shergottites, nakhlites, and chassignites, named after the original type meteorite of each class, Shergotty, Nakhla, and Chassigny. The Martian meteorites are collectively known as the SNC (pronounced "snick") meteorites. In early 1994 yet another new meteorite type, recovered from the surface of the Antarctic ice cap, was identified as a fragment of the crust of Mars. This meteorite does not belong to any of the established SNC families. The ejection of SNC meteorites logically requires the loss of atmosphere at the same time that the meteorites are accelerated to escape velocity.

Rocks are of course a minor accessory to the loss of gases from the explosion fireball. Only the very edge of the expanding hemispherical impact fireball even touches the ground. For each kilogram of rock successfully accelerated to escape velocity, millions of kilograms of gas must have been lost into space from the Martian atmosphere.

So easy is it to erode atmosphere from Mars that there must have been serious loss of gases over geological time scales. At MIT in 1982, Hampton Watkins and I tried to estimate how much of the Martian atmosphere might be lost by such "impact erosion." To our surprise, we found that more than 90 percent of the mass of the early atmosphere of Mars could have been lost in this way. More recent detailed computer models of the impact process on Mars by Jay Melosh and Ann Vickery confirm not only that impacts can blast meteorites off the planet, but also that the early atmosphere might have been dozens of times as massive as the remanent that survives today. The tenuous nature of the atmosphere of Mars remarked upon in chapter 7 therefore is plausibly explained by the erosive effects of impacts.

Impact explosions upon Venus have nearly the same escape velocity to overcome as on Earth (10.2 vs. 11.2 kilometers per second), but the atmosphere of Venus is some thirty times denser than Earth's. The atmosphere therefore helps muffle the explosion, and makes erosion of the atmosphere of Venus very difficult. "Pollution" of the atmosphere by trapped cometary gases, discussed in chapter 9, is far more important than atmospheric erosion on Venus. Also, because of its very dense atmosphere, Venus can disrupt much larger bodies, and have much more powerful atmospheric explosions, than we have on Earth. Thus the best present-day evidence of "atmospheric cratering" in the solar system is the presence of air-blast scars on the surface of Venus. If we were to examine the floor of Earth's oceans after a major oceanic impact, we would probably see scoured features similar to those seen today at the base of Venus's atmosphere. But these are delicate features: although much more durable than craters in ocean water, these scars on the abyssal plains of the oceans will be covered by layer after layer of sediment, soon to become unrecognizable. Identifying such scars may prove as difficult as linking tsunami deposits unambiguously to impacts.

In general, both shock waves from airbursts and tsunami waves from ocean impacts may present serious hazards to populated areas. We now know enough about the nature of these hazards to search for and assess records of human casualties caused by cometary and asteroidal materials. That shall be our purpose in chapter 13.

13

EFFECTS ON HUMAN POPULATIONS

Parkersburg, West Virginia, March 10, 1897. A meteor burst over the town of New Martinsville yesterday. The noise of the explosion resembled the shock of a heavy artillery salute. . . . When the meteor exploded the pieces flew in all directions, like a volcanic upheaval, and solid walls were pierced by the fragments. David Leisure was knocked down by the force of the air caused by the rapidity with which the body passed before it broke. The blow rendered him unconscious. One horse had its head crushed and nearly torn from the trunk by a fragment of the meteor, and another horse in the next stall was discovered to be stone deaf.

THE NEW YORK TIMES
March 11, 1897

That arrogant attitude of incredulity, which rejects the facts without any attempt to investigate them, is sometimes almost more injurious than unquestioning credulity.

ALEXANDER VON HUMBOLDT
Kosmos, c. 1840

It is easy to find statements in the press, and even in scientific journals, asserting that "no one in recorded history has ever been killed by a meteorite," that "only one person has ever been struck by a meteorite," that "meteorites are not hot when they fall, and have never been known to start a fire," and so on. Such authoritative statements do not invite skepticism. However, my research for this book took me to many primary sources, including many hundreds of eyewitness accounts of meteorite falls and violent atmospheric explosions. I have read hundreds of reports in English, and translated many more from French, German, Russian, Italian, Latin, and, with help from Leo Masursky, Chinese. I have concluded that the above generalizations might better agree with the eyewitness reports if they were changed as follows: "No one in recorded history has ever been killed by a meteorite in the presence of a meteoriticist and a medical doctor"; "Only one person in recorded history who was struck by a meteorite was interviewed by a twentieth-century American reporter";

and "Meteorites have never been observed to start a fire in the presence of a meteoriticist and a fire marshal." Another useful conclusion from my reading is that the reviewers who make sweeping negative conclusions usually do not cite any of the primary publications in which the eyewitnesses describe their experiences, and give no evidence of having read them. They thus willingly assume the role of Defenders of the Faith and Guardians of the Truth, practicing that arrogant attitude of incredulity so heartily despised by von Humboldt.

The attitude of many modern experts toward damage, injuries, and deaths inflicted by meteorites is astonishingly similar to the attitude of the intellectual elite of Europe toward meteorite falls before 1800. The sources of the eyewitness reports, then as now, are of course mostly from rural areas. They are mostly nonprofessional. They are sometimes, especially in underdeveloped countries, illiterate. And finally, they are commonly laughed at by snotty ignoramuses who have evidently never been outside at night. Over and over again we find the eyewitness reports followed a few days later by a letter to the editor, in which some Dr. Defensor Veritatis points out that the fire the farmer put out could not possibly have been started by the meteorite he picked up in the same place because it is "well known that meteorites are cold when they fall, and that frost often grows on their surfaces." As for reports of injuries or fatalities reported in other languages, or (even worse) in other centuries, clearly the observers must have been little more than superstitious, credulous, and impressionable children. It is easier to scoff at the credentials of the witness than to accommodate the observation within the expert's theory.

Now, what should we believe when we read an account like the one given at the beginning of this chapter? Here we have a very realistic account of an airburst and the fall of fast-moving meteorite fragments, of damage to property and livestock, and injury to a man. We do not know the witness personally; it is no longer an option to interview him and subject him to a battery of tests. But the reporter who wrote the piece quite possibly knew the man; he could examine the damage for himself and interview the neighbors to establish the probity of the witness. Could he have made up such a realistic story? Certainly neither the witness, Mr. David Leisure, nor the reporter, could be expected to have a firm grasp of the physics of airbursts, aerodynamics, aerial fragmentation, and meteorite impact. Their *interpretations* of the event are therefore worthless. But their *observations* of the event are priceless. Note that I use the word *their*, not *his*. This is because a skilled reporter will question his sources carefully to distinguish observations from interpretations, and pass on the grain while discarding the chaff. (I recognize that several contemporary publications thrive by turning this process on its head; however, I have not used them as sources.)

Several dozen of the reports I have read contain statements that read something like this: "A spectacular fireball coursed the sky over Xville last night, alarming the citizens with a hideous buzzing noise and the brilliant light of its passage. The bolide, which was about the size of a hat, flew along some sixty feet above the ground before passing out of sight behind Farmer Y's barn. As it fell there was a loud detonation, like the discharge of a cannon, followed by a rumbling as of distant musketry. Many persons in the vicinity were awakened by the blast, and some were thrown from their beds. Windows were shattered and ceilings cracked in several nearby towns. The blast was heard as far as sixty miles away. A strong smell of brimstone was noticed after the fall. A diligent search of the field behind Farmer Y's barn turned up nothing unusual."

This typical report contains many descriptions of direct *sensory* phenomena. It also contains two blatant interpretations: that the fireball was "the size of a hat," and that it flew "some sixty feet above the ground." The observer has no means at his disposal to measure either the size or the distance of the object. Indeed, when we find other reports of the same sighting from communities twenty miles away, we find a close agreement on almost everything. There will generally be only one small area in which the smell of sulfur was noted; the other observer will agree that the bolide was "the size of a hat," and even that it was "traveling sixty feet above the ground," but when the direction and position of the object is correlated between several such observers, it can be calculated quite securely that the fireball was 120 kilometers above the ground when first sighted, and the terminal explosion was, say, 10 kilometers up. The calculated size of the fireball may be several hundred meters in diameter.

Why "sixty feet up," then, and why "the size of a hat"? When the reporter thinks to ask these questions the answer usually comes out like this: "It had to be sixty feet up because it cleared the barn, and the barn is fifty feet high." The astonishing frequency with which people compare bolides to hats probably follows from this first judgment: if the bolide really were sixty feet up, then it would have to have been about the size of a hat. The only other size comparison I have found repeated is "a fireball the size of a barn." (One is tempted to conclude that hats and barns are of great importance in rural areas.) Apparently the very high angular speeds of these bodies convince the naive observer that they must be very near by. If they were aware that the bolide was traveling thirty kilometers per second instead of thirty feet per second, their impressions of its distance and size would be radically altered.

My conclusion from reading many such reports is that the actual observations reported are in fact highly reliable, whatever the educational attainments of the observer. Interpretations can generally be distinguished from the observations. Educated observers tend to propose interpretations that are more complex and, overall, more informative and useful than those of the untrained, but this advantage is offset by the tendency of the sophisticated to let their theories color their observations: the educated are rarely content to report what they saw, but feel obliged to explain it, and often to edit their observational account to accord with their theory. But both peasant and scholar know a fire or a dead body when they see one.

Well, then, what about meteorites that are hot enough to set fires when they fall? Let's read a few eyewitness accounts. The sources for these records are listed in the table at the end of this chapter.

In the year 476, in I-hsi and Chin-ling, China, "thundering chariots" that were "like granite" fell to the ground. "All of the vegetation was scorched to death."

The *Philosophical Magazine*, series I, volume 16 (1808), page 294, reports a meteorite that fell in 1620 in the Punjab, India. The meteorite burned the grass on which it fell. The meteorite, which was made of iron, was later made into a dagger, a knife, and two sabers.

A 1700 meteorite fall in Jamaica is reported in *Philosophical Transactions*, 30, 837 (1718) in a letter by Mr. Henry Barham. He witnessed a brilliant fireball and "great Blaze": "I observed many holes in the Ground, one in the middle of the Bigness of a Man's Skull. . . . the green Grass was perfectly burnt near the Holes, and a strong Smell of Sulphur remain'd thereabouts for a good while after."

From the *Mémoires de l'Académie des Sciences* (1759): "On June 13 of this year 1759, about 9 o'clock in the evening, the heavens being clear and calm, with a cold wind blowing from the north, the priest of the village of Captieux (two leagues from Bazas) noticed in the sky a column of fire, which seemed to direct itself from the east to the meridian; but soon his view was obscured by some trees. Upon reaching his home, barely had he gone to bed, but he heard the cry of "Fire!"; his brother immediately ran to the stable, where the fire appeared; the flames having already filled every part of the stable, and having disappeared as quickly, he found four horses which had just been killed, without any mark of burning, and that all the stall litter had been consumed by the fire; then he noticed a smell of sulfur so strong that it seemed suffocating; he had great difficulty recovering. Meanwhile, the loft floor of the stable had not been burned at any point, only two holes of three or four thumbs (inches) diameter were found; but all the roofing was aflame, and it was necessary to sacrifice the stable to save the house."

Volume 1 of the *Mémoires de l'Académie de Dijon* reports that "in the night of 11 to 12 October 1761, a house in Chamblan, a half league from Seurre (near Bourgogne) was set fire because of the fall of a meteor."

A fireball appeared over Baton Rouge, Louisiana, on April 5, 1800, with a brightness "little short of the effect of sun-beams." A "considerable degree of heat was felt by those who saw it . . . and in a few seconds after a tremendous crash, similar to that of the largest piece of ordnance, causing at the same time a very sensible earthquake." Where it fell there was "scorched vegetation . . . the surface of the earth broken up." The event was reported in the *Philosophical Magazine*, series I, volume 11, page 191 (1801).

On July 4, 1803, in East Norton, England, a meteorite fell and burned the grass it landed on, according to the *Philosophical Magazine*, series I, volume 16, p. 191.

On October 30, 1801, in Suffolk, England, the "dwelling-house of Mr. Woodroffe, miller, near Horringer-mill, Suffolk, was set on fire by a meteor, and entirely consumed, together with a stable adjoining." This was reported in *The Times* on November 3, 1801.

The *Annuaire* tells us in 1836 of a fireball that fell from the sky and set fire to a barn at Belley, in the Department de l'Ain, France, on November 13, 1835. Note the date!

In the *Comptes Rendus de l'Académie des Sciences*, volume 12, p. 1196 (1841), we find a reference to *Annales des Voyages 15*, concerning a meteorite that fell in the Chiloé Archipelago on the coast of Chile and started a fire.

In *The New York Times* of December 7, 1907, we read of a meteorite that fell in Bellefontaine, Ohio, starting a fire that destroyed a house. On April 27, 1910, there is another story of a giant meteor that burst in the mountains in Mexico, starting a brushfire. A similar event on July 15, 1921, started a fire in the Berkshire Hills of western Massachusetts. On February 24, 1933, in Stratford, Texas, a bright fireball dropped a four-pound iron meteorite which burned the grass where it landed. A spectacular, widely observed meteorite fall in Newfoundland on October 19, 1936, set a fishing boat afire.

An even more recent account of this phenomenon concerns a reported meteorite fall in Kirkland, Washington, on January 17, 1955. This event, described in an article in *Meteoritics*, volume 2, page 56, involved two iron meteorites of about 110 grams each that penetrated the dome of an amateur astronomer's observatory, smashed a clock on the wall, and came to rest inside the dome. Luther L. Hawthorne, the seventy-year-old owner of the observatory, reported that he heard a loud bang, looked out the kitchen window, and saw two holes in the dome. A few moments later, he saw smoke issuing from the holes and immediately telephoned the fire department. He then ran out and opened the dome. He related that one of the fragments had come to rest on a row of reference books and had set fire to them. The other fragment was found on the floor. He doused the small blaze before the fire department arrived on the scene. The story was reported in the *Seattle Times*, but seems not to have been picked up by the wire services or brought to the immediate attention of meteoriticists. The meteorites were much abused before being subjected to professional inspection: their fusion crusts were sandpapered away to expose the metallic interior, and crude attempts were made to remove samples.

Several aspects of this story engendered skepticism. First, the meteorites crossed the dome traveling at an angle of only twenty-seven degrees to the horizon. One would normally expect such small bodies to reach the ground falling more nearly vertically. Second, evidence of thermal alteration of the exposed surface is lacking. This may be a consequence of either prolonged weathering of the surface (which rusts away the heated surface layer) or to intentional abrasion to expose the metallic interior. Finally, the meteorite is a medium octahedrite similar to many of the Canyon Diablo irons from Meteor Crater, which are readily available on the commercial market. In light of the similarity to Canyon Diablo, the absence of a thermally altered layer suggests either fraud (the irons are weathered pieces of Canyon Diablo that have had their rusted surfaces removed) or ignorance (the irons are authentic fresh falls, but the fusion crust was removed for misguided aesthetic reasons). Oh yes, one other criticism that was raised: it couldn't be an authentic fall because real meteorites don't start fires! Thus the jury remains out on this event. Could the impact of the meteorite with the dome have deflected its path toward the horizontal? Did Mr. Hawthorne stand to gain from a hoax? Why did the only meteoriticist who visited the scene conclude that it was probably authentic?

Finally, a report published in 1969 in the *Journal of the British Astronomical Association*, volume 79, page 475, tells of the fall of an iron meteorite on November 14, 1968, in Alandrola, Portugal. The meteorite was described as being incandescent after landing on the ground.

IT is interesting that these reports of fires, whenever they are explicit, concern iron meteorites, not the vastly more common stony meteorites. In general, stony meteorites are scoured clean of their melt by aerodynamic forces, leaving only a very thin layer of thermally altered surface no more than a millimeter thick. Such a thin heated layer will cool quickly. Indeed, many reports of the fall of stones assert that the meteorite felt slightly warm moments after it fell, only to grow frost on its surface a few minutes later. This is exactly what we expect for stones in orbit about the Sun near Earth's orbit. Without an atmospheric greenhouse effect to help keep their heat in, stones are colder than the mean surface temperature of Earth, and commonly below the freezing point in their interiors. The only meteorites with high enough thermal conductivity to store up some of their entry heat and be hot at the time of fall are irons and stony irons.

Reports of persons being injured or killed by meteorites are uncommon. The account in Joshua 10:11, dating from about 1420 B.C., may be the oldest: "And it came to pass, as they fled from before Israel, and were in the going down to Beth-horon, that the Lord cast down great stones from heaven upon them into Azekah, and they died: they were more which died with hailstones than they whom the children of Israel slew with the sword." The use of both "great stones" and "hailstones" in this passage is interesting. These "credulous, ignorant" people evidently accepted that large stones and lethal blocks of ice (or stones that quickly became coated with ice!) could fall from the sky. This is the same passage in which the Sun and Moon are commanded by Joshua to stand still in the heavens, after which it remained bright for "a whole day," which, in order to be meaningful in this context, must mean at least twenty-four (and therefore thirty-six) hours of illumination, not twelve. The most direct interpretation is that one entire night was about as bright as day. If we recall reports from England on the night after the Tunguska fall, or imagine the night sky dominated by a brilliant passing

· ·

comet, the juxtaposition of these two phenomena makes some physical sense.

Kevin Yau, Paul Weissman, and Don Yeomans of the Jet Propulsion Laboratory have recently reported on a search for Chinese records of meteorite falls that caused damage, death, or injury. Their fascinating discoveries include a number of claims of human fatalities. For example, one report from China describes a fall of stones from the sky upon an army camp on January 14, A.D. 616. The stones destroyed several siege towers and "crushed to death more than ten people."

One astonishing event with near-fatal consequences is recounted by Einhard, the biographer of Charlemagne, in his *Vita Caroli Magni*, written about A.D. 815: "One day, on the expedition [A.D. 810] which he led into farthest Saxony, against Godefried, the King of the Danes, just before the rising of the Sun, as he was setting out from his camp, as he went along the road he suddenly saw (a meteor in) the form of a gigantic light sliding down the sky, dashing from right to left through a clear sky. As everyone came to a halt and stared, wondering at the meaning of this portent, the horse which he [Charlemagne] was riding suddenly threw down its head and fell, hurling him to the ground so violently that the clasp of his cloak was broken and his sword-belt torn asunder."

There are two reported incidents of people being killed by meteorites in Italy. The first, on September 14, 1511, alleges the death by celestial stoning of several birds, a sheep, and a monk, a feat which evidently requires a number of stones. (Or could this be the origin of the "seven at one blow" story?) The second, sometime between 1633 and 1664, also involved a monk, a Franciscan friar in Milano, who was reported killed by a meteorite that struck him on the leg. In the absence of eyewitness accounts and authenticating evidence, it would be difficult in either of these cases to get a conviction in a modern court before a jury of meteoriticists and criminologists. However, people in 1511 were no less intelligent than we are now. They simply didn't know that preserving the meteoritic "smoking bullet" and the notarized coroner's report for five centuries would be necessary in order to have anyone believe them. Those who were eyewitnesses to these events might be astonished to find that we do not believe that stones fell from the sky and struck a man. They might be angered to hear that we are willing to assume that they were deluded or lying, without having any evidence to suggest that they were unreliable witnesses. For my part, I believe that we should take these reports seriously, while at the same time regarding them as unproven by the evidence available to us.

The seventeenth-century event described above, which had been universally discounted by modern meteoriticists, took on a new life in 1985 when research by Marco Cavagna and Massimo Vicentini uncovered an account written by the physician who examined the body of the Milanese monk. The doctor, Paolo Maria Terzago, described the injury in detail. The stone, striking the friar in the thigh at high speed, severed his femoral artery. He quickly bled to death from the injury. I suppose this doctor's report will be ignored because no Ph.D. meteoriticist or socially acceptable Guardian of the Truth was present to authenticate the stone that fell from the sky as a "true" meteorite.

Yau and colleagues find three different though mutually consistent records from between the years 1321 and 1368, of a particular fall of iron meteorites that had tragic consequences. One of these reports reads: "It rained iron to the east of the Erh River. Houses and hill tops were damaged. People and animals who encountered them were mostly killed. It was known as the " 'iron rain.' " A separate incident in 1369 reads, "General Hsu Tai-szu led an army to the Ho-t'ao area. At a place named Wu-t'ung-shu, one day at noon time, there was a large star which fell into the river. It started a fire which spread to the river bank. Some soldiers from the camp were wounded by it."

In February or March of 1490, three different sources describe another remarkable event. According to Yau, et al.: "Stones fell like rain in the Ch'ing-yang district (Shansi Province). The larger ones were 4 to 5 catties (about three pounds) in weight and the smaller ones were 2 to 3 catties (about two pounds). They struck dead more than ten thousand people. All of the people in the city fled to other places." Another source describes the stones as ranging from the size of goose eggs down to the size of water chestnuts. A third source speaks of "tens of thousands" of deaths.

Of course such a rain of stones could be terribly dangerous, but is it really possible that tens of thousands of small stones should fall from the sky at the same time? Isn't this an obvious example of the grossest exaggeration? In fact, there is ample precedent for "blizzards" of stones in the historical record. The famous L'Aigle fall of L chondrites on April 26, 1803, dropped over three thousand stones on the French countryside. A fall of over one thousand L-group (low-iron ordinary chondrite) stones was seen at Knyahinya, Russia, on

. . . .

June 9, 1866. On February 3, 1882, another fall of many thousand L-chondrite stones occurred near Mocs, Romania. The Forest City, Iowa, fall of May 2, 1890, dropped over two thousand H-group (high-iron ordinary chondrite) stones. The Holbrook, Arizona, fall on July 19, 1912, contained over fourteen thousand L-chondrite stones, and that at Dokachi, in present-day Bangladesh, on October 22, 1903, also contained many thousand H-chondrite stones. Other falls in which thousands of stones were documented include those at Malotas, Argentina (June 22, 1931; H chondrite), Pantar, Philippines (June 16, 1938; H chondrite), and Pueblito de Allende, Mexico (February 8, 1969; CV3 chondrite). The Sikhote-Alin fall of February 12, 1947, rained thousands of iron meteorites, some of which were large enough to produce explosion craters over a hundred meters in diameter.

The largest documented meteorite fall took place near Pultusk, Poland, on January 30, 1868. On that occasion, over one hundred thousand H-group chondritic meteorites fell in a rural area in winter, when no one was in the fields. If this fall had struck a city, vast numbers of casualties would have resulted. Perhaps the 1490 fall in Shansi Province was similar to the Pultusk shower. Interestingly, and soberingly, a shower of stony meteorites just missed the city of Jilin, in Jilin Province of China (Manchuria), on March 8, 1976. The elliptical area within which several hundred stones fell to the ground is offset to the north of the city by an amount equal to the width of the ellipse. This was surely a close call! The largest Jilin stone weighed in at well over one thousand kilograms.

The Chinese historical sources that tell us about the Shansi disaster also describe an event in 1639 in which "a large stone fell suddenly unto a small market street. It destroyed several tens of houses. The number of people killed amounted to several tens."

Another story from the mid-1600s tells of two sailors who were killed by the fall of stones on shipboard in the Indian Ocean, en route from Japan to Sicily. So improbable and bizarre is the notion that stones should fall out of the sky in mid-ocean and kill men that there was surely a strong incentive to suppress such stories at their source. Indeed, the twentieth is the first century in which we have recorded evidence, authenticated by meteoriticists and medical doctors alike, of a human being struck by a meteorite. Why should we, of all people, decide that this story is "not credible"?

According to an article in the *Philosophical Magazine*, Series I, volume 16, page 293, a meteorite fell near Barbotan and Agen in Gascony, France, on July 24, 1790. The meteorite crushed a cottage, killing a farmer and some cattle.

On January 16, 1825, in Oriang (Malwate), India, a man was reported killed and a woman injured by a meteorite fall. On January 23, 1870, in the village of Nedagolla, India, a man was stunned by a meteorite.

From China comes a report that on June 30, 1874, at Chin-kuei Shan, Ming-tung Li, China, a huge stone fell in the midst of a thunderstorm. A cottage was crushed, and a child within was killed. And on September 5, 1907, according to Yau, a stone fell at Hsin-p'ai in Weng-li, China: "The whole of Wan Teng-kuei's family was crushed to death."

Among the eyewitness accounts of the June 30, 1908, Tunguska explosion collected by Russian researchers, several pertain to Tungus reindeer herders who were in the area of destruction. Of the approximately twenty people who were within fifty kilometers of ground zero, it appears that all were at least slightly injured. Several reindeer herds totaling well over one thousand head were "burned to ashes." Vasily Dzhenkoul reported the loss of seven hundred reindeer and all of his herd dogs. An old man named Vasily was thrown twelve meters against a tree, breaking his arm. He later died of his injuries. Ivan Yerineev was thrown to the ground and bit off his tongue. Ivan Dzhenkoul lost over two hundred reindeer. An old hunter, Lyuburman, died of shock. Ivan Akenov was thrown to the ground and knocked unconscious. S. Dronov lost his entire reindeer herd and fell injured in a remote area of the taiga, where he lay unconscious for two days. The nearest glass windows, seventy kilometers from ground zero in the trading station of Vanavara, were blown out by the blast wave, fortunately striking nobody.

According to the December 8, 1929, report in that notorious supermarket tabloid, *The New York Times*, a meteorite struck a wedding party in the town of Zvezvan, Yugoslavia, killing one person. The *New York Times* also reported on May 16, 1946, that a fall of meteorites had struck Santa Ana, Nuevo León, Mexico, destroying many houses and injuring twenty-eight persons.

We have scarcely considered injuries and near-misses of human targets, but such reports exist. On June 16, 1794, in Siena, Italy, a meteorite struck a child's hat as it sat on her head. Fortunately, the child was uninjured. On February 27, 1827, in Mhow, India, a man was struck on the arm by a meteoritic stone that shattered a nearby

. . . .

tree. In Braunau, Bohemia (now the Czech Republic), on July 14, 1847, a thirty-seven-pound iron smashed through the roof of a house into a room where three children were sleeping. There were no significant injuries.

On November 19, 1881, a man was reported injured by a falling meteorite at Grossliebenthal, which was then a town in Russia.

Next came the orthodox example, the "only person ever struck by a meteorite." On November 28, 1954, Mrs. Annie Hodges of Syla- cauga, Alabama, was struck by an eight-pound stony meteorite that pierced the roof of her house, demolished her radio, and ricocheted into her leg. This event, and this event alone, has been certified by both socially acceptable meteoriticists and a medical doctor, and therefore is "credible."

Finally, we find an account in *Meteoritics* describing the fall of the Mbale meteorite in Uganda on August 14, 1992. The area covered by the fall, about twenty square kilometers, includes the village of Mbale. Several roofs were damaged by falling stones. The largest of the forty-eight recovered pieces weighs 27.4 kilograms, more than adequate for lethality. A boy was hit on the head by a tiny 3.6-gram fragment which fortunately had dissipated its energy by passing through a banana tree. The press has been strangely silent about this event, which is vouched for by professional meteoriticists.

Near misses are certainly much less newsworthy than injuries or deaths. Nonetheless, we do have records of many near misses from relatively modern sources. On January 1, 1869, in Hessle, Sweden, a falling meteorite missed a man by a few meters. The same thing happened on January 3, 1877, in Warrenton, Missouri, and again on January 21, 1877, in De Cewsville, Ontario.

On July 19, 1912, a spectacular shower of over 14,000 stony me- teorites fell in a very sparsely populated area near Holbrook, Ari- zona, missing one man by a few meters. On February 2, 1922, in Baldwyn, Mississippi, a man was missed by 3 meters; similarly, on May 30, 1922, in Nagai, Japan. On July 6, 1924, in Johnstown, Colorado, a man was missed by one meter. On August 8, 1933, in Sioux County, Nebraska, and again on August 11, 1935, in Briggsdale, Colorado, and on April 6, 1942, in Pollen, Norway, other close calls were re- corded. On March 3, 1953, in Pecklesheim, in the Federal Republic of Germany, a person was missed by several meters.

On June 15, 1984, a meteorite missed a man in Nantong, China, by several meters. Two sunbathers soaking up the healing rays on a beach in Binningup, Australia, were missed by about five meters by a definitely nonhealing meteorite on September 30, 1984. A 9.5-kilogram meteorite smashed through a house roof in La Criolla, Argentina, on January 6, 1985, destroying a door and missing a woman by only two meters. On July 2, 1990, in Masvingo, Zimbabwe, a man was missed by 5 meters. Also, on August 31, 1991, in Noblesville, Indiana, a falling meteorite traveling at an angle of about thirty degrees to the horizon (!) missed two boys by 3.5 meters. Perhaps it's time to reassess the Kirkland, Washington, "hoax" of 1955, in which the closely similar entry angle was used as an argument against the authenticity of the reported fall of two similar-sized irons!

Livestock also are not immune to meteorites. We have noted the fourteenth-century Chinese report of the death of animals in a meteorite fall and the 1511 incident in Cremona when several birds and a sheep were killed in the presence of the unfortunate monk; also, the 1790 fall in Gascony in which the farmer was killed also caused the death of some cattle. In addition, on December 11, 1836, in Macao, Brazil, a fall damaged several homes and killed several oxen. On May 1, 1860, in New Concord, Ohio, a colt was struck and killed. On March 11, 1897, in New Martinsville, West Virginia (the story quoted at the beginning of this chapter), a man was knocked out and a horse was killed. On June 28, 1911, a dog in Nakhla, Egypt, was killed by an SNC meteorite, the first known example (H. G. Wells notwithstanding) of any terrestrial lifeform being killed by something from Mars. And on June 24, 1938, in Chicora, Pennsylvania, a cow was struck and injured by a falling stone. Table 2, summarizing a number of reports of damage, injury, death, and close calls is included at the end of this chapter.

What should we conclude about the hazard from meteorite falls? The most important conclusion is that meteorite falls constitute an utterly negligible hazard compared to a single large multimegaton airburst such as the Tunguska explosion. In the face of this judgment, which is statistically unassailable, it makes little or no difference whether one person is killed by meteorites each century, or one hundred people are killed each year. The reason that stories of individual fires, deaths, injuries, and near misses are valuable is that they put the cosmic threat in human terms: we can relate to a reindeer hunter in Siberia who is thrown through the wall of his hut, the Argentine housewife who sees her kitchen smashed before her eyes, the student in New York who finds her car with a head-sized hole blasted through its trunk, the French priest whose stable is burned and whose horses are killed by an errant fireball. We relate much less successfully to the image of a city of 10 million people, shredded with glass shrapnel and incinerated by a firestorm. We may indeed find the concept horrifying, but we do not relate to it on a practical level. Perhaps the main reason that strategic weapons control has taken so long is that nuclear megadeath is beyond our abilities to understand or to imagine; it is literally unthinkable. Let us not make the same mistake with impact megadeath.

We cannot leave this chapter with the conviction that the record cited here is complete, or even representative. There have been many powerful selection effects militating against reporting of meteorite-caused casualties and property damage. Some of these selection effects are demographic: most of Earth has had, until very recently, a very low population density. Almost all of the populated areas were very rural, with both a low population density and a distinct shortage of literate observers. Probably most of the lethal events that have occurred have killed their only eyewitnesses. Other negative selection effects are fundamentally sociological; for a variety of reasons, many of the literate kept no historical records; then there was the contempt or disregard of the lettered elite for the quaint tales and assumed credulity of the peasantry; the official rejection of meteorites before 1800 in the West because of their offensiveness to pure reason; the domination of the print media by the most entrenched Guardians of Traditional Wisdom, and so on. Victor Clube and Bill Napier go even further: they see evidence of a de facto conspiracy to silence any who dared rock the preternaturally stable boat of uniformitarian geology, biology, and sociology. But one need not subscribe to conspiracy theories to see that various forms of censorship and discrimination must have profoundly deterred the publication and preservation of these reports.

Thanks to the advances described in chapters 1 through 13, we can do a better job of describing one thousand years of the impact history of Earth from computer modeling than we can from historical records alone. To do so requires an integration of all we know about the numbers, orbits, and physical properties of bodies in near-Earth space with the most recent understanding of the chemistry and physics of impacts. If we are willing to take on this difficult task, we may succeed in seeing, if not the actual history of the next century, at least a range of possible futures. Such an understanding is an essential prerequisite of intelligent planning for survival.

. . . .

PROPERTY DAMAGE, INJURIES, AND DEATHS CAUSED BY METEORITE FALLS

Date	Place	Source	Event
c. 1420 B.C.	Israel	Joshua 10:11	Lethal meteorite fell
A.D. 476	I-hsi and Chin-ling, China	Yau, et al. (1993)	"Thundering chariots" "like granite" fell to ground; vegetation was scorched
06/25/588	China	Yau, et al. (1993)	"Red-colored object" fell with "noise like thunder" into furnace; exploded; burned several houses
01/14/616	China	Yau, et al. (1993)	10 deaths reported in China from meteorite shower; siege towers destroyed
764	Nara, Japan	Met. **1**, 300 (1963)	Meteorite strikes house
810	Upper Saxony	*Vita Caroli Magni*	Charlemagne's horse startled by meteor; threw him to ground
1064	Ch'ang-chou, China	Yau, et al. (1993)	Daytime fireball, meteorite fell; fences burned
1321–1368	O-chia District, China	Yau, et al. (1993)	Iron rain killed people, animals, damaged houses
1369	Ho-t'ao, China	Yau, et al. (1993)	"Large star" fell, started fire, soldiers injured
02-03/?/1490	Ch'ing-yang, Shansi, China	Yau, et al. (1993)	Stones fell like rain; over 10,000 killed
10-11/?/1504	China	Yau, et al. (1993)	"Large star" fell with "noise like thunder; garden burned
09/14/1511	Cremona, Lombardy, Italy		Monk and several birds and a sheep killed
1620	Punjab, India	*Phil. Mag.* I **16**, 294 (1808)	Hot iron fell, burned grass. Made into dagger, knife, 2 sabers
1639	China	Yau, et al. (1993)	Large stone fell in market; tens killed; tens of houses destroyed
1633–1664	Milano, Italy		Monk reported killed by meteorite
1647–1654	Indian Ocean		Two sailors reported killed on ship en route from Japan to Sicily
08-09/?/1661	China	Yau, et al. (1993)	Meteorite smashed through roof; no injuries
11/07/1670	China	Yau, et al. (1993)	Meteorite fell, broke roof beam of house

Date	Place	Source	Event
06/13/1759	Captieux, France	Mém. Acad. Sci. (1759)	Stable struck and burned by meteorite
10/11/1761	Chamblan, France	Mém. Acad. Dijon 1C.R. Acad. Sci. **7**, 76	House struck and burned by meteorite
07/24/1790	Barbotan and Agen Gasc. France	Phil. Mag. I **16**, 293 (1808)	Meteorite crushed cottage, killed farmer and some cattle
06/16/1794	Siena, Italy		Child's hat hit; child uninjured
12/19/1798	Benares, India		Building struck
10/30/1801	Suffolk, England	Times 11/03/1801 p. 3d	"Dwelling-house of Mr. Wood-roffe, miller, near Horringer-mill, Suffolk, was set on fire by a me-teor, and entirely consumed, to-gether with a stable adjoining."
07/04/1803	E. Norton, England	Phil. Mag. I, **11**, 191 (1803)	White Bull public house struck, chimney knocked down, grass burned; flight nearly horizontal
12/13/1803	Massing, Czech.		Building struck
11/10/1823	Waseda, Japan	Met. **1**, 300 (1963)	Meteorite struck house
01/16/1825	Oriang (Mal-wate), India		Man reported killed, woman in-jured by meteorite fall
02/27/1827	Mhow, India	Phil. Mag. IV **25**, 447	Man struck on arm, tree broken
11/13/1835	Belley, Dept de l'Ain, France	Annuaire (1836)	Fireball set fire to barn
12/11/1836	Macao, Brazil		Several homes damaged, several oxen killed by meteorite
1841	Chiloé Archipelago, Chile	C.R.A.S. **12**, 1196	Fire caused by meteorite fall
05-06/?/1845	Ch'ang-shou, Szechwan, China	Yau, et al. (1993)	Stone meteorite damaged over 100 tombs
07/14/1847	Braunau, Bohemia		37-lb. iron smashed through roof into room where three children were sleeping; no serious injuries
10/17/1850	Szu-mao, China	Yau, et al. (1993)	Meteorite fell through roof of house
12/09/1858	Ausson, France		Building hit
05/01/1860	New Concord, Ohio		Colt struck and killed
08/08/1868	Pillistfer, Estonia		Building struck

Date	Place	Source	Event
01/01/1869	Hessle, Sweden		Man missed by a few meters
01/23/1870	Nedagolla, India		Man stunned by meteorite
12/07/1872	Banbury, England	Nature **7**, 112	Fireball felled trees, wall
06/30/1874	Chin-kuei Shan, Ming-tung Li, China	Yau, et al. (1993)	Thunderstorm; huge stone fell, crushed cottage, killed child
02/16/1876	Judesegeri, India		Water tank struck
01/03/1877	Warrenton, Missouri		Man missed by a few meters
01/21/1877	De Cewsville, Ontario		Man missed by a few meters
01/14/1879	Newtown, Indiana	Paducah Daily News	Leonidas Grover reported killed in bed (hoax?)
01/31/1879	Dun-le-Poelier, France	C. Flammarion	Farmer reported killed by meteorite
11-12/?/1879	Huang-hsiang, China	Yau, et al. (1993)	Rain of stones; many houses damaged; sulfur smell
11/19/1881	Grossliebenthal, Russia		Man reported injured by meteorite
11/22/1893	Zabrodii, Russia		Building struck
02/10/1896	Madrid, Spain		Explosion; windows smashed, wall felled
03/11/1897	New Martinsville, West Virginia	NYT 03/12/87 1:4	Man knocked out, horse killed; walls pierced
11/04/1906	Diep River, South Africa		Building struck
09/05/1907	Hsin-p'ai Wei, Weng-li, China	Yau, et al. (1993)	Stone fell; whole family crushed to death
12/07/1907	Bellefontaine, Ohio	NYT 12/08/07 1:4	Meteorite started fire, destroyed house
06/30/1908	Tunguska Valley, Siberia		Two reportedly killed, many injured by Tunguska blast
05/29/1909	Shepard, Texas	NYT 05/30/09 1:6	Meteorite dropped through house
04/27/1910	Mexico	NYT 04/28/10	"Giant meteor" burst, fell in mountains, started forest fire
06/16/1911	Kilbourn, Wisconsin	NYT 08/08/32 17:6	Meteorite struck barn
06/28/1911	Nakhla, Egypt		Dog struck and killed by meteorite

Date	Place	Source	Event
07/19/1912	Holbrook, Arizona		Building struck; 14,000 stones fell; man missed by a few meters
01/09/1914	Western France	NYT 01/10/14 1:7	Meteor explosions broke windows
11/22/1914	Batavia, New York	NYT 11/23/14 1:8	Meteorites damaged farm
01/18/1916	Baxter, Missouri		Building struck
12/03/1917	Strathmore, Scotland		Building struck
06/30/1918	Richardton, North Dakota		Building struck
07/15/1921	Berkshire Hills, Massachusetts	NYT 07/15/21 15:2	"Meteor" started fire in Berkshires
12/21/1921	Beirut, Syria		Building hit
02/02/1922	Baldwyn, Mississippi		Man missed by 3 meters
04/24/1922	Barnegat, New Jersey	NYT 04/25/22 1:2	Rocked buidings, shattered windows, clouds of noxious gas
05/30/1922	Nagai, Japan		Person missed by several meters
07/06/1924	Johnstown, Colorado		Man missed by 1 meter
04/28/1927	Aba, Japan		Girl struck and injured by dubious meteorite
12/08/1929	Zvezvan, Yugoslavia	NYT 12/09/29 III 1:2	Meteor hit bridal party, killed 1
06/10/1931	Malinta, Ohio	NYT 06/11/31 3:4	Blast, crater, smell of sulfur, windows broken in farmhouse; 4 telephone poles snapped, wires downed
09/08/1931	Hagerstown, Maryland	NYT 09/09/31 14:2	Meteorite crashes through roof
08/04/1932	Sao Christovaõ, Brazil	NYT 08/04/32 6:5	Fall unroofed warehouse
08/10/1932	Archie, Missouri	NYT 08/13/32 17:6	Homestead struck, person missed by <1 m
02/24/1933	Stratford, Texas	NYT 03/25/33 17:1	Bright fireball, 4-lb metallic mass; grass burned
08/08/1933	Sioux County, Nebraska		Man missed by a few meters
02/16/1934	Texas	NYT 02/17/34 32:3	Pilot swerved to "avoid crash"
09/28/1934	California	NYT 09/29/34 1:3	Pilot "escaped" shower

Date	Place	Source	Event
08/11/1935	Briggsdale, Colorado	NYT 08/11/35 21:2	Man narrowly missed by meteorite
03/14/1936	Red Bank, New Jersey	NYT 03/17/36 23:3	Meteorite through shed roof
04/02/1936	Yurtuk, USSR		Building struck
10/19/1936	Newfoundland	NYT 10/20/36 27:7	Fisherman's boat set fire by meteorite
03/31/1938	Kasamatsu, Japan	Met. **1**, 300 (1963)	Meteorite pierces roof of ship
06/16/1938	Pantar, Philippines		Several buildings struck
06/24/1938	Chicora, Pennsylvania		A cow struck and injured
09/29/1938	Benld, Illinois		Garage and car struck by 4-lb stone
07/10/1941	Black Moshannon Park, Pennsylvania		Person missed by 1 meter
04/06/1942	Pollen, Norway		Person missed by 1 meter
05/16/1946	Santa Ana, Nuevo León, Mexico	NYT 05/17/46 9:4	Meteorite destroyed many houses, injured 28
11/30/1946	Colford, Gloucestershire, UK	NYT 11/31/46 7:6	Telephones knocked out, boy knocked off bicycle
02/12/1947	Sikhote Alin, near Vladivostok	NYT 04/29/47 14:3	Iron meteorites fell; cratering
09/21/1949	Beddgelert, Wales		Building struck
11/20/1949	Kochi, Japan		Hot meteoritic stone entered house through window
05/23/1950	Madhipura, India		Building struck
09/20/1950	Murray, Kentucky	NYT 09/21/50 33:7	Several buildings struck
12/10/1950	St. Louis, Missouri		Car struck
03/03/1953	Pecklesheim, FRG		Person missed by several meters
01/07/1954	Dieppe, France	NYT 01/09/54 2:6	Meteorite, blinding explosion, smashed windows
11/28/1954	Sylacauga, Alabama	NYT 11/19/54 86:2 and Met. **1**, 125 (1963)	Mrs. Annie Hodges struck by 4-kg meteorite that crashed through roof, destroyed radio.
01/17/1955	Kirkland, Washington	Met. **2**, 56 (1964)	2 irons broke through amateur astronomer's observatory; one set on fire

Date	Place	Source	Event
02/29/1956	Centerville, S. Dakota		Building hit
10/13/1959	Hamlet, Indiana		Building hit
02/23/1961	Ras Tanura, Saudi Arabia		Loading dock struck
09/06/1961	Bells, Texas	Met. **2**, 67 (1964)	Meteorite struck roof of house
04/26/1962	Kiel, FRG		Building hit
12/24/1965	Barwell, England		Two buildings and a car struck
07/11/1967	Denver, Colorado		Building struck
04/12/1968	Schenectady, New York	Met. **4**, 171 (1968)	House hit
04/25/1969	Bovedy, Northern Ireland		Building hit
08/07/1969	Andreevka, USSR		Building hit
09/16/1969	Suchy Dul, Czechoslovakia		Building hit
09/28/1969	Murchison, Australia		Building hit
04/08/1971	Wethersfield, Connecticut		House struck by meteorite
08/02/1971	Haverö, Finland		Building hit
03/15/1973	San Juan Capistrano, California		Building hit
10/27/1973	Canon City, Colorado		Building hit
08/18/1974	Naragh, Iran		Building hit
01/31/1977	Louisville, Kentucky		Three buildings and a car struck
05/13/1981	Salem, Oregon		Building hit
11/08/1982	Wethersfield, Connecticut	JRAS Canada **85**, 263 NYT 11/10/82 1:1 and 01/02/83 I 33.5	Pierced roof of house
06/15/1984	Nantong, PRC		Man missed by 7 meters
06/30/1984	Aomori, Japan		Building struck
08/22/1984	Tomiya, Japan		Two buildings hit
09/30/1984	Binningup, Australia		2 sunbathers missed by 5 meters

. . . .

Date	Place	Source	Event
12/05/1984	Cuneo, Italy		Strong explosion, blinding flash; windows broken; daytime fireball "bright as Sun"
12/10/1984	Claxton, Georgia		Mailbox destroyed by meteorite
01/06/1985	La Criolla, Argentina		Farmhouse roof pierced, door smashed. 9.5-kg stone missed woman by 2 meters
07/29/1986	Kokubunji, Japan		Several buildings hit
03/01/1988	Trebbin, GDR		Greenhouse struck by meteorite
05/18/1988	Torino, Italy		Building struck
06/12/1989	Opotiki, New Zealand		Building hit
08/15/1989	Sixiangkou, PRC		Building hit
04/07/1990	Enschede, Netherlands	New Sci., 6/9/90, 37	House hit by believed fragment of Midas
07/02/1990	Masvingo, Zimbabwe		Person missed by 5 meters
08/31/1991	Noblesville, Indiana	Sky & Telescope 4/92	Meteorite fall missed 2 boys by 3.5 m
08/14/1992	Mbole, Uganda	Met. 29, 246	48 stones fell; roofs damaged, boy struck on head
10/09/1992	Peekskill, New York	S.E.A.N.	Car trunk, floor pierced by meteorite

C.R.A.S.—Comptes Rendus de l'Académie des Sciences
NYT—*The New York Times*
S.E.A.N.—Smithsonian Earth and Atmospheric Notices
Yau, et al. (1993)—A preprint circulated by Kevin Yau, Paul Weissman, and Don Yeomans of the Jet Propulsion Laboratory
J.R.A.S.—Journal of the Royal Academy of Science

Incidents given without citation were widely reported by many different publications.

14

THE FIERY RAIN:
SIMULATIONS BY COMPUTER

ASTEROID STRIKES OFF CHINA COAST: MILLIONS FEARED DEAD. Singapore, Vladivostok Shaken by Blast. Sydney, June 28. Towering tidal waves, thrown up by the impact of a small asteroid in the Philippine Sea, last night struck deep into the interior of China and devastated coastal areas of Japan, Korea, Taiwan, the Philippines, Indonesia, and many West-Pacific islands, according to fragmentary reports from sources in Australia and eastern Russia.

All communications with this area have been lost. The tragedy is now being pieced together from a variety of reports from outside the immediate impact area. The story began to unfold around 3 P.M. Sydney time yesterday, when an astronomer at the Anglo-Australian Observatory at Siding Spring, Australia, found, on a photograph taken the previous night, the trail of a previously unknown small asteroid-like body with a trail that suggested that it was closer to Earth than the Moon and approaching Earth. The first eyewitness reports came at 11:49 P.M. Tokyo time last night, when two airline flight crews radioed reports of a brilliant fireball and explosion about 1,200 miles southeast of Taipei, Taiwan. The first of these, a Qantas flight en route from Tokyo to Sydney, estimated the explosion as hundreds of miles ahead of it on its route southward over the Philippine Sea. The second report, from a Tokyo-bound Japan Air Lines 747 out of Manila, described a spectacular "column of fire" that "rose over the eastern horizon and turned night into day for several minutes." Communications with both flights were quickly lost, and their fate is unknown.

A massive communications blackout, attributed by industry sources to electromagnetic pulse effects and massive disruption of the ionosphere, has enveloped the area from the north coast of Australia to the northern tip of Japan. Sources in the U.S. Air Force Space Command report that a Ballistic Missile Early Warning satellite, stationed in synchronous orbit 24,000 miles above the Western Pacific, observed a "tremendous explosion" near 20°N and 135°E, near the tiny island of Perece Vela. The Air Force announcement says that the explosion had a power of thousands of megatons, so large that it "could not have been any known type of nuclear weapon." Sources in NASA have verified that a number of research satellites in low orbits around the Earth went silent simultaneously.

The Far Eastern Command of the Russian Air Forces in Vladivostok reports the cessation of all radio and television transmissions from the south. Several brilliant fireballs in the Western Pacific were reported by the crew of the Novy Mir space station in the minutes preceding the main impact, at which time all contact with the station was lost.

Weather satellite images of the area taken from synchronous orbit show an immense circular area of dense clouds above the impact site. A tidal wave front was seen to overrun Okinawa and run up onto the east coasts of the Philippines and Taiwan. The tidal wave, estimated to be hundreds of yards in height, reached the coast of China just before press time this morning, devastating the coast from Fuchou to Shanghai. At press time, the tidal wave had reached the Yangtze River valley, one of the most densely populated areas in the world. The wave overran the Ryukyu Islands south of Japan, and had reportedly reached the city of Kago-shima on the southern tip of Japan's Kyushu Island. Traveling at a speed of about 500 miles per hour, the tidal wave was at that time only an hour away from Hong Kong, Seoul, and Tokyo.

Scientists at Los Alamos National Laboratory warn that the effect of the impact will not be limited to the Western Pacific area. Their timetable for the arrival of the tidal wave listed Osaka and Kobe, Japan, at 5:30 A.M. eastern time, Tokyo at 5:55, Seoul at 6:10, Vladivostok at 7:15, Singapore at 8 A.M., Honolulu and Sydney at 11 A.M., Auckland at noon, Los Angeles and San Francisco at 4 P.M., Vancouver and Seattle at 5 P.M., and Lima, Peru, and the Panama Canal around 10 P.M. eastern time. Preliminary estimates suggest that the tidal wave may run up to a height of 30 to 50 feet when it reaches the North American coast.

Intermittent satellite communications with Singapore portray a deadly struggle for survival since the appearance of two brilliant fireballs at sunset and the later arrival of the earth shock from the main impact. With no high land accessible, there has been a stampede of hundreds of thousands of people, many of them armed, to the airport. Military discipline has reportedly collapsed, with armed military personnel leading the dash to the airport. No reports on the state of warnings and evacuation plans in the cities of southern Japan have been received. The impact, occurring shortly before midnight local time, apparently knocked out all communications before warning could be given. The flash from the explosion may have awakened many people, and the impact, equivalent to an earthquake of Richter magnitude 8, was felt widely outside the blast area. The tidal wave reached the southern tip of Japan at about 2 A.M. local time. . . .

May we never read such a tale of horror in our newspapers. Certainly, such events can occur. Impact explosions of 1,000 megatons or more occur with average intervals of about eight thousand to ten thousand years. This seems extremely long to a person who expects to live eighty years: a hundred lifetimes! But we know of a 350-megaton event in Argentina that could not be more than a few thousand years old, and we know that about three oceanic impacts of similar size must occur for each land impact. We also know that recorded human history extends back only some six thousand years. It is simple to calculate the probability that a 1,000-megaton impact event has occurred in those six thousand years of historical time: the odds *favor* an impact of that size by a 2:1 margin!

But there is another way of looking at the same story. What actually happened eight thousand to ten thousand years ago to end the hunter-gatherer chapter in human history? Quite suddenly, agriculture became common, specialized occupations arose, and cities appeared. Writing was invented, giving rise to record-keeping and literacy. And the earliest human records all record stories of floods that devastated civilization. The Babylonian Epic of Gilgamesh, the Greek Flood of Deucalion, and the Noachic flood are merely the most familiar examples. The study of the literature of catastrophes even has a name: it is called *eschatology*. Modern biblical scholars seem generally intent upon tracing which eschatological writings may have been imported from other cultures, completely ignoring the most important question: What actually happened? Central Asian and Amerind traditions describe the emergence of dry land from beneath a global ocean (a peculiar concept to arise among Plains Indians if they "invented" the story!). What *did* happen then? Was the clock of human history reset to zero by an event (or more than one) that devastated civilization?

Deciphering the true history of that era is perhaps no longer possible, but projecting the future is another matter. We already have in our possession a wealth of data on the hazards of impacts. We know with certainty that stones of all sizes, from centimeters (chapter 1) to tens of meters (chapter 2), bombard the Earth in modern times. We know that many bodies, like the Tunguska bolide, explode with extreme violence in the atmosphere and leave no lasting geological evidence (chapter 3). We know from the effects of nuclear weapons tests and from the bitter experience of Hiroshima and Nagasaki how large explosions affect humans and human constructions (chapter 4). We know from spacecraft missions to the Moon and Mars (chapter 5) what the average rate of cratering has been in the inner solar system in recent times. Astronomers, building upon technological breakthroughs in detector and computer technology arising from military research and the civilian space program, have refined their methods so that they can discover and count Earth-crossing asteroids and comets with sizes down to a few meters (chapter 6). We have learned from the study of Mars (chapter 7) that global dust storms can seriously affect climate, and from the study of the end-Cretaceous clay layer (chapter 8) that Earth was shrouded with a massive dust storm mingled with meteorite dust, soot, and acid rain 65 million years ago.

In the last few years the *Magellan* radar-mapping spacecraft has found powerful proof of the bombardment of Venus by comets and asteroids and the selective destruction of smaller and weaker impactors by explosions in the dense atmosphere (chapter 9), giving us a valuable check on our theories of explosions in Earth's atmosphere. Earth-based radar studies have meanwhile found polar ice caps on Mercury, providing yet another check on our theories of the injection and removal of atmospheric gases by impacts (chapter 10). We have seen that the evidence from any single planet is inadequate to test or confirm the overall theory, but, by great good fortune, the technologically advanced nations of Earth have sent forth spacecraft to explore other planets. These missions had already given us deep insight into the effects of water vapor and chlorine compounds on the stability of the ozone layer; but now they also inform us about another hazard capable of not only removing the ozone layer, but devastating civilization and wreaking terrible havoc on the entire biosphere.

Using our recent studies of comets and asteroids, it has become clear that average impact rates do not tell the whole story. Impact "storms" must occur (chapter 11). Most of the impactors strike the ocean (chapter 12), where some may deposit their own content of water and other ices, as on Mercury and Venus, or, in sufficiently large impacts, even blast off part of Earth's atmosphere into space. But all oceanic impacts throw up massive tidal waves that can devastate coastal regions without leaving any distinctive signature that says, "this was an impact event." Human casualties are possible from any event that drops kilogram-sized meteorites, and many such events have been reported (chapter 13), only to be dismissed by meteorite experts who demand absurdly high standards of proof. But clearly the real hazard lies with larger, rarer bodies. Even a modest aerial explosion like the fifteen-megaton Tunguska event would utterly devastate a modern city. The growth of global population in the last few centuries, the urbanization of human culture into glass boxes, and the vast increase in population along the seacoasts, all conspire to increase the risk from airbursts and tsunamis enormously.

What are the likely events of the twenty-first century? What can we expect to occur in our lifetimes? This question can now be answered *statistically* by means of the process of computer simulation. All the available evidence on nearby bodies in space is folded into a statistical model of the size, composition and strength, abundance, and orbits of the near-Earth asteroid and comet population. All the evidence on the effects of giant explosions from studies of nuclear weapons tests, cratering, gas injection, airbursts, fire ignition, shock-wave chemistry, acid rain, and atmospheric erosion on Earth and other planets is included in statistical form. All presently understood hazards to life and property, insofar as we have discovered them, are also included. We can then run the model for a period of a century (a natural human time scale) to see what might happen. I have chosen the twentieth century as the setting for these runs because we know the population, population density, and technological abilities of Earth for that time period. The twentieth century provides a vehicle for more meaningful exploration of the influence such cosmic events would have had upon the world.

At our present state of knowledge, statistical calculations of what *could* happen are justified and achievable. If we knew more, however, then we could calculate what really *will* happen. If we had, instead of only a statistical description of a small part of the Earth-crossing comet and asteroid fluxes, a nearly complete catalog of all the bodies in near-Earth space and their precise orbits, masses, and physical properties, then we could literally calculate the fate of each body in the catalog, giving the dates at which they strike one planet or another or are ejected from the solar system. But since our data are still incomplete, the best we can presently do is to generate statistically reliable computer models of the typical behavior of the orbiting swarm. From cratering rate data, for example, we can say how many kilometer-sized bodies will strike North America per 100 million years. But if we had our orbit catalog completed, we could calculate the actual dates of those impacts.

The first and most fundamental discovery from these computer models is, not surprisingly, that the impact history is subject to enormous statistical variations. Even without formally including clustering of impacts due to the existence of orbital "families" and fragmentation of large comets, great differences are found between different one-century runs.

The fairest kind of prediction, therefore, must not only convey one likely outcome, but also give an appreciation of the range of variation that can be expected from one century to the next. For this reason we will present a narrative account of ten consecutive one-century runs, just as they came off the computer. In the interests of saving time and space, only the fate of the most massive, obvious, and hazardous objects in each year will be considered. We shall sensationalize the results of these ten runs by giving them the dramatic names "Scenario A" through "Scenario J." Each is as valid and probable a version of the twentieth century as the one recorded in our newspaper files.

SCENARIO A

In 1901 an unobserved and unreported forty-kiloton airburst of a hard stony (achondritic) meteorite occurs five kilometers above a remote ocean area near the coast of Queen Maud Land in Antarctica. In fact, in an average year there is one atmospheric explosion with a yield of one hundred kilotons or more. The large majority occur in such remote areas, or so high in the atmosphere, that they are not observed. Even if observed, the witnesses may see only a flash of light in the distance, or hear the "rumble of distant thunder" coming from the open oceans. Thus even those that are observed are often not recognized. Only "reportable" events and especially interesting unobserved happenings will be mentioned hereafter.

In 1902 a 2,970-metric-ton asteroidal fragment of ordinary chondrite material grazes the atmosphere over the South Atlantic at a speed of 29 kilometers per second. It flies nearly horizontally through the upper atmosphere along a trajectory that is less curved than the surface of the Earth, passing as close as 112 kilometers above sea level. The flare of its passage is as bright as −10.5 magnitude, comparable to a crescent Moon. The fireball is visible for about half a minute before the object exits from the atmosphere with its original speed virtually undiminished. The probability that this event would be witnessed, recognized, and reported is only about 0.01.

In 1906, and again in 1909, brilliant fireballs of about −20 magnitude (one hundred times as bright as a full Moon) are seen over Calgary, Alberta, and Kano, Nigeria. Both fireballs terminate with high-altitude explosions that shake the ground and rattle windows, but do no damage. Events of this sort are reported locally, but seldom picked up by national and international news media. They are usually called "mysterious explosions" or "earthquakes" when they occur in daytime and "exploding meteors" when they occur on clear nights. One or two such easily observed events occur each year in major metropolitan areas in the United States alone.

In 1914 a fragment of a dead long-period comet explodes thirty-seven kilometers above the middle of the Indian Ocean, with an explosive yield of 10.9 megatons. This is to be the greatest aerial explosion of the century, but, because of its high altitude and remote location, it is unobserved. Perhaps some fishermen or a crewman on a freighter sees the flash or hears the boom at a distance; in any event, it is not reported.

From 1918 to 1923 there are four more brilliant fireballs over rural areas on land. Two occur in daytime, and are not seen. One, on a cloudy night, is noted only by the sound of its explosion. The fourth, falling early on a clear evening, is seen and heard by hundreds of observers near Cartagena, Colombia.

Three megaton-class explosions occur between 1933 and 1940. The 1933 blast, a 1.3-megaton explosion of an ordinary chondrite body, occurs at only 11-km altitude. The other two, in 1937 and 1940, blow up at 35-km (1.7 megatons yield from weak carbonaceous chondrite material) and 37-km altitude (6.9 megatons from the explosion of weak cometary debris). All three events occur far out over the oceans and are unobserved. In 1955 another fragment of cometary debris blows up in a 1.2-megaton explosion at an altitude of 36 kilometers over a sparsely populated rural area in northern Iran. The daytime blast is heard over a roughly circular area some 70 kilometers in diameter. It rattles windows and pops a few doors open, but does no damage. Several witnesses report a flash of light in the sky and a transient dark gray cloud.

In 1969 and 1971 two more megaton-class explosions occur over the oceans. Neither is observed.

In 1974 an iron object weighing eighteen hundred metric tons enters the atmosphere in the mid-Pacific at an entry speed of only 11.6 kilometers per second. It crashes into the ocean unobserved, but not far from busy shipping lanes, some 3,200 kilometers from the nearest coast, striking the sea surface at 10.7 kilometers per second. It explodes upon impact with the force of twenty-five kilotons of high explosive, slightly more than the Hiroshima atomic bomb. A weak tidal wave radiates out from the impact, too low and too broad to be noticed by ship crewmen among the wind-driven ocean swells, but sufficient to run up to a height of half a meter and wash high onto hundreds of beaches in North America, Hawaii, and Japan. Over eleven hundred people are caught in the "freak wave," and many are drowned, taken unawares by its sudden onset. No connection with an impact event is ever made.

A megaton explosion occurs in 1985 at an altitude of thirty-nine kilometers over the North Atlantic. No effects are observed on the surface, but infrared detectors aboard a Soviet military satellite observe the event. It is at first tentatively identified as a clandestine nuclear weapons test, but correlation of data from other satellites shows that the electromagnetic pulse (EMP) signature of a megaton nuclear explosion is definitely not present. The data are stamped "top secret" and archived.

In 1988 and 1990 two large bodies (7,400 and 1,900 metric tons) pass through the upper atmosphere and "skip out" at speeds well above escape velocity. They approach to minimum altitudes of 108 and 77 km, respectively, and are not observed.

Finally, in late 1999, another skipout event occurs just as the century is drawing to a close. This event lights up the skies across Western Europe with an evening fireball as bright as a full Moon. Millions of people observe the fireball as it courses across the sky for forty-four seconds over a path of eleven hundred kilometers. Several tracking stations of the European network photograph the fireball and confirm that its orbit carries it back out into interplanetary space.

The scorecard for the century: over one hundred people killed and one thousand more injured by an impact-induced tsunami, but no connection is made between the (unobserved) impact and the casualties. Experts report "no hazard from meteorites."

The scenarios that follow shall include only events that were witnessed and caused injuries, damage, or death.

SCENARIO B

In 1929 a piece of achondrite material (hard stone) weighing seventeen hundred metric tons strikes the ground in the early morning hours in a wheat-farming community north of Saskatoon, Saskatchewan. The twenty-five kiloton explosion levels a village, killing 381 people. Fires set in outlying homes by the radiant heat of the fireball claim another 63 lives.

In 1965 a forty-seven-hundred-metric-ton iron projectile with a diameter of about 9 meters lands in a rural area in the state of Bengal in eastern India. The forty-one kiloton explosion excavates a crater 280 meters in diameter. Every structure in an area 3.5 kilometers in diameter is demolished, killing 162 people. Fires are ignited in the rubble, but cause no additional fatalities.

SCORECARD: This century featured an extraordinary number of

cometary high-altitude airbursts, improbably numerous (thirteen!) megaton explosions, and an improbable targeting of the airspace over continents by the bombarding bodies (thirty-eight out of one hundred of the annual biggest events occurred over land, compared to 29 percent for the long-term average). Total deaths due to impact: 606, in two fatal events. Popular wisdom accepts impacts as a real hazard.

SCENARIO C

An iron of about twenty-two hundred metric tons, entering at a slightly steeper angle (eleven degrees below the horizontal, rather than nine) and a slightly lower speed than the 1913 skipout event, impacts the ground intact with an explosive power of twenty-three kilotons, blasting a crater 240 meters in diameter. Debris ejected at supersonic speed from the impact crater cuts a swath through a small town in China, and the impact blast wave destroys every building over an area 2.9 kilometers in diameter. The death toll is 628.

In 1963 a slow-moving 9,900-metric-ton fragment of an achondritic near-Earth asteroid explodes with an explosive yield of 144 kilotons just four kilometers above the ground in Spain, damaging many buildings and shattering windows with the force of the blast, and igniting fires over an area more than two kilometers in diameter. Nineteen people are killed.

SCORECARD: Two lethal events in the century, totaling 647 deaths. All large explosions are safely out at sea. As before, a few small, dense, strong bodies dominate the death toll, while most large explosions are at high altitudes and therefore frightening but harmless. The impact hazard is again widely recognized.

SCENARIO D

In 1946 a 25,000-metric-ton achondritic fireball explodes at 4:00 A.M. local time at a height of eleven kilometers above Fergana, Uzbekistan. The 1.2-megaton blast damages buildings over an area several kilometers in diameter, searing the area with intense heat and setting thousands of fires. The fires burn out of control, killing 4,146. Over 20,000 residents are awakened by a brilliant flash of light and heat to find their city in flames. An "earthquake" is reported by the survivors. Several metric tons of meteorite fragments are mixed in with the debris of two thousand burned-out and collapsed buildings, where they are indistinguishable from scorched and blackened fragments of structural brick and rock. Since the event occurred in a known earthquake-prone region, the possibility of an airburst is not considered. Seismic monitoring of the area is so poor in 1946 that no evidence bearing on the cause of the disaster is collected.

In 1954 a piece of a pallasitic (strong stony-iron) asteroid explodes with thirty-one kilotons of force just 1.4 kilometers above the ocean off the coast of Peru. A fishing boat is ignited by the intense radiation from the fireball. Two crewmen are severely lacerated by shattered window glass. The third crewman on the boat is burned to death when the boat's fuel tank explodes. The boat burns to the waterline and sinks, leaving no surviving witnesses or evidence.

A spectacular 1.7-megaton explosion in 1977, some thirty-seven kilometers above Great Falls, Montana, causes a local panic and floods telephone switchboards with calls to police and fire officials. Simultaneous reports of a large explosion from DSP satellites in geosynchronous orbit and Air Force personnel at Malmstrom Air Force Base persuade Norad to move to Defcon 2. The "missile farms" in the area, clusters of Minuteman ICBM silos and control centers, are kept sealed off for several hours as Norad analyzes satellite data on the explosion and determines that it was not an attack. What would have happened if this had been an impact event that produced a one-hundred-kiloton surface explosion within the ICBM missile farm?

SCORECARD: There were six megaton explosions in the century, of which the five biggest caused no damage or injuries. There was one mass fatality due to an airburst in an earthquake-prone region. There were no observations to link this disaster to its true cause. A second event resulted in three fatalities with no surviving witnesses. Conclusion: no impact hazard.

SCENARIO E

The sole large explosion over land is a 2.8-megaton explosion seventeen kilometers above the outskirts of Kyoto in 1922. The radiation from the fireball ignites thousands of fires over an area thirteen kilometers in diameter, triggering a firestorm in which 8,591 people die. The weak blast wave from the high-altitude explosion, arriving thirty-five seconds after the ignition of the fires, brings down many burning buildings and hinders rescue attempts but is ineffectual at blowing out the fire. The event is extremely well documented: the aerial explosion is observed directly by hundreds of survivors, and over a million hear the explosion. The cause is not, however, understood for several decades.

In 1979, a low-altitude forty-three-kiloton airburst over the Mediterranean Sea south of Rome destroys an Alitalia airliner en route to Palermo, killing 159 passengers and crew. Observers in the Naples area witness the event. After a thorough investigation, complicated by a false claim by the Red Brigade that they had blown up the plane as a political act, the cause of the disaster is discovered. Many distorted fragments of meteoritic iron are later dredged up from the area where the wreckage fell.

SCORECARD: An extraordinary cluster of events in 1978–1983, including the aviation disaster and two observed megaton-sized airbursts, calls attention to the hazard from comet and asteroid falls. Theorists attempting to understand the Italian disaster discover the mechanism responsible for the Kyoto firestorm. Impacts are recognized as a serious threat in 1980, but only at the cost of 8,759 lives in two deadly events. An intensive asteroid search, named "Spaceguard," is begun immediately.

SCENARIO F

In 1930 a forty-two-hundred-metric-ton iron asteroid crashes into the ocean off the west coast of Ireland during heavily overcast, rainy weather with a fifty-four-kiloton explosion. Fifty fishermen are killed instantaneously by the blast. No witness of the event survives.

An achondritic stone of about the same size explodes 2 km above the sea off the north coast of Java in 1939. The eighty-seven-kiloton explosion capsizes and sinks several small ships in the busy shipping lanes near Djakarta, and 572 people are lost. A number of survivors are pulled from the water and tell of the enormous explosion that destroyed their ships. None actually saw the entry of the bolide, but all saw the flash of the explosion.

In 1964 a 1.2-megaton explosion high over Amman, Jordan, visible in the evening skies from Cairo to Damascus, starts wildfire rumors that Israel has carried out a nuclear attack on its neighbor. Frenzied fighting erupts between Palestinians and Israelis, only to fade out when it becomes clear that no damage has been done by the explosion. Both the king of Jordan and the prime minister of Israel call for calm. Many in the region, Muslim, Christian, and Jew, interpret the explosion as a "celestial portent." Some see it as a warning to make peace; others interpret it as a signal for the start of a crusade or jihad.

In 1993 a rare, strong achondritic stone explodes 1.2 kilometers above the ground near Santa Rosa, Argentina. The blast wave knocks down trees and levels buildings out to 1.8 kilometers from ground zero. The death toll from the initial blast is 704. Many people who survive the blast are trapped in the rubble of their homes, and 106 more are killed by fire before rescuers can reach them.

SCORECARD: This was a truly spectacular century, with over 30 million witnesses of large airbursts and three events that caused mass death and destruction. In this scenario, the public demands action to predict impacts and protect against their effects. It is difficult to convince them that they just had a run of bad luck. (After all, they can't know what happened in our other scenarios!) They would be even more alarmed if they knew that the seventy-five largest explosions of the century resulted in not a single casualty, just by the luck of where they fell, their strength, and their composition.

SCENARIO G

In early 1910, a very slow-moving sixty-eight-hundred-metric-ton pallasitic stony-iron asteroid falls 1,400 km from the African coast, near St. Helena island in the South Atlantic. A weak tidal wave runs ashore from Dakar to South Africa, running up to a height of about 60 centimeters on most beaches. The tidal surge is focused and amplified as it enters the mouths of the Congo and Niger Rivers, causing the sudden inundation at midnight of many riverside villages by 1 to 2 meters of water. Over five thousand people, mostly the elderly and young children from inland villages, are drowned. The wave is much higher and more violent at St. Helena, but no one was on the beach on the side facing the tidal wave when it arrived. Damage is light, and there are no deaths or injuries because the residents live well above sea level. The tsunami is variously described as a "freak of nature," a "storm surge driven by distant weather disturbances," or a "tidal wave from an underwater earthquake in the South Atlantic."

SCORECARD: Popular wisdom is unanimous: impacts are not a hazard. Not a single fatality or injury was attributed to impact in this century; but 5,019 are dead from an impact-induced tsunami whose nature was not understood. Ten megaton-size explosions generate no fatalities. The pattern grows clearer: it is the rare strong and slow bodies that kill. Impacts on land are severe short-range hazards, but impacts in the ocean are dangerous even at very long ranges. The total energy of the impact is much less important than the location of the resulting explosion: strong impactors penetrate deeper into the atmosphere, closer to where the people are. Slow-moving (<15 km/second) bodies also penetrate deeper before the aerodynamic pressure (which increases with the square of their velocity) can crush them.

SCENARIO H

The event of the century occurs in 1928. A fireball almost as bright as the Sun appears over the English Channel, racing westward at 27 kilometers per second. The fireball is visible in the predawn sky from much of southern England, and in France as far southeast as Paris. In London the overcast nighttime skies are lit from above as by the rising of the Sun. An ordinary chondrite body weighing 1.2 million metric tons slices into the atmosphere at a shallow angle, shedding an incandescent plume of rock vapor 350 kilometers long, and explodes 22 kilometers above the busy North Atlantic sea-lanes just off Land's End, England. The explosion has the power of 102 megatons of TNT, seven times the yield of the Tunguska projectile in our own century, larger than any nuclear weapon ever detonated by the United States or the Soviet Union. The searing heat of the explosion illuminates the surface of the ocean at ground zero for a fraction of a second with the light of a thousand Suns. A thousand square kilometers of the ocean surface boil fiercely, and an oil tanker bursts like a kernel of popcorn, gushing flames and smoke. All exposed crewmen on nearby ships are charred instantly by the heat. Those below decks hear a deafening groaning and popping as the immense thermal stress of the fireball radiation twists and bends the hull and deck plates, firing rivets about like bullets; then sudden silence and darkness as the electrical systems of their ships are mangled and shorted out. A string of passenger ships en route between North America and Europe, following the main trans-Atlantic shipping lanes, are seared by the heat of the fireball; their wooden lifeboats and the ropes holding them burst into flames; all windows on the port side admit actinic light with the intensity of a cutting laser. The brilliance of light even inside these windows is deadly: window shades, upholstery, and clothing smoulder briefly and burst into flames. Their crews, watching the brilliant fireball approaching them almost head-on, are at first dazzled by the light, but the vastly brighter flare of the final explosion literally burns out their eyes. Ships out of New York, Halifax, Newport News, Cherbourg, Southampton, Philadelphia, Baltimore, Copenhagen, Quebec, New Orleans, Amsterdam,

Oslo, and Providence fill with smoke as they careen on, unpiloted, into hell.

A full minute later the shock wave slams into the sea. The blast wave shocks an area forty-four kilometers in diameter, instantly shredding five cargo ships and a Cunard liner and sending them to the bottom. The tanker, with its one hundred thousand barrels of blazing oil, is blown out like a candle on a child's birthday cake. The blast wave rolls the tanker over for three complete revolutions, tearing the hull open in a hundred places and wrenching the stern free of the rest of the hulk. A hot blast of wind traveling at nearly the speed of sound arrives behind the shock front, whipping oil and steam across the ocean surface at over five hundred miles per hour. The blast, weakening as it expands in a sphere forty kilometers across, strikes a smoldering passenger liner broadside, ripping off its funnels and radio masts. The high profile of the ship spells its doom: it rolls far to port, hangs there a few moments with a forty-five-degree list, and then capsizes. Its lifeboats are stripped off by the blast and disappear in the raging winds.

Several days pass before the toll can be counted: British, Norwegian, French, and German liners, an American-registered tanker, five merchantmen sunk and fourteen more torched, reduced to mute hulks. A severely damaged Royal Navy destroyer burns through the day with exploding ordnance and great sudden flares of burning bunker oil. The number of dead exceeds ten thousand . . . and the burn victims continue to die.

A Lloyds inquiry reaches the astonishing conclusion that a natural explosion of unimaginable fury, a giant meteor or meteorite disintegrating high in the upper atmosphere, was the culprit. Public awareness of the impact hazard is achieved at terrible cost.

In 1955 an American carrier task force in the Sea of Japan is struck by another aerial explosion, this one with an explosive force of 2.2 megatons. An 80,000-metric-ton ordinary chondrite body explodes at a height of thirteen kilometers, scorching the aircraft carrier, a heavy cruiser, and three accompanying destroyers. The carrier, conducting flight operations at the time, has nineteen fully fueled aircraft on its flight deck when the blast strikes. The fuel tanks on the planes rupture and burst into flame. The starboard catapult launches a blazing fighter through the first sheet of flames. In a moment the flight deck is an inferno of aviation fuel. The captain and rear admiral, viewing the aircraft-launching operations from the island, are blinded by the flash. Thinking they are under nuclear attack, the officer of the day sounds general quarters, then calls for emergency fire-fighting details to report to the flight deck. Several jet fighters in flight near the carrier, orbiting to await the arrival of their comrades who have not yet been launched, burst into flames and plummet into the sea, one striking like a kamikaze into the bridge of the cruiser. Ammunition from the burning aircraft on the flight deck starts cooking off, spraying the deck with shrapnel. A fully fueled Skyray fighter, just loaded onto the aft port elevator from the hangar deck, erupts into flame as crewmen scramble to move other fuel-fat aircraft back from the flames. A magazine of fifty-caliber ammo cooks off on the Skyray, ripping through the hangar deck, shredding the fuel tanks on a dozen other planes. Then the 500-pound bombs start to go.

The blast wave from the sky arrives with barely enough strength to blow out the flames on the flight deck. For a moment there is silence and a sudden surge of hope, but as the wind subsides, red-hot fragments of aircraft reignite the aviation fuel. The carrier, mortally wounded, lies dead in the water, drifting broadside to the wind, while the cruiser sails in a great arc, unpiloted. Two destroyers come alongside the flattop to take off crew, but it is too late. The flames and bomb fragments breach the magazine. With a tremendous explosion, the guts of the carrier are torn out. Bombs in the magazine detonate in chain reaction in a split second. Accidental nuclear explosions cannot occur; the bombs are designed so they cannot be exploded by any chance event. But the bomb casings and high explosive charges in nuclear weapons cannot withstand fire and explosive shock. The high-explosive triggers of dozens of atomic bombs detonate like a string of firecrackers, showering radioactive debris from the nuclear weapons over all three ships. Flakes of burning plutonium fall as gently as snow on firefighters and rescuers.

The carrier, glowing through the night with its immense burden of radioactivity, is too hot to handle. It is scuttled at dawn, sunk by a salvo of torpedoes from one of its brood of destroyers a hundred kilometers off the coast of Japan. The news cannot be suppressed despite the most stringent efforts to maintain secrecy. Finally President Truman acknowledges the accident. Thousands of leftist Japanese students riot in protest.

The cause of the disaster emerges months later from a classified military inquiry. The new president, General Eisenhower, elected on a platform that emphasized distrust of the military-industrial complex, decides to make the full findings public. The report of the review board concludes unequivocally that the accident was caused by a high-altitude aerial explosion of an asteroidal body. Chairman Khrushchev of the Central Committeee of the Communist Party of the USSR, a charismatic and outspoken apparatchik with the manner of a peasant and the mind of a chess grand master, holds this explanation up for public ridicule. Ignoring all the published evidence, which he claims is concocted, he says that there was an accidental nuclear explosion on board the carrier, and condemns the United States for militarism, warmongering, and incompetence in the handling of nuclear weapons. He calls for UN sanctions against the United States.

The final death toll from all direct causes is 5,452, not counting those killed in rioting and the sack of American embassies in a dozen countries.

In 1989, a second billion-metric-ton ordinary chondrite arrives at Earth, following an almost identical orbit and arriving with nearly the same speed as the 1928 object. It falls precisely on the sixty-first anniversary of the 1928 impactor that killed ten thousand in the Atlantic Ocean off Land's End. This time we are not so lucky.

It enters the atmosphere fifteen kilometers south of Orleans, France, detonating at an altitude of nineteen kilometers with an explosive force of eighty-three megatons. Fires are ignited as far as seventy-eight kilometers from the blast. An area of nineteen hundred square kilometers is devastated by the blast wave. Nearly every building within twenty-five kilometers of ground zero is flattened. Out to a distance of seventy-five kilometers, windows are blown in and shards of glass are accelerated to high speeds. The shock wave, when it reaches the ground, blows out most of the flames within thirty kilometers, but many hot spots remain inside the outer circle of flames. The center of the city is devastated. Modern glass-walled buildings are traversed and swept empty by the shock wave; deadly shards of glass, mingled with the furniture, partitions, and occupants of the buildings are blown out the lee side of the building at high speeds, to fall as a thick layer of impenetrable debris several stories high blocking all the streets. About 40,000 people are killed by the blast wave. The gas tanks of cars ignited by the flash continue to burn beneath the layer of combustible rubble, setting off a growing conflagration that quickly escalates into a full-scale firestorm. The tower of flame over Orleans draws in winds from all sides. As the fire spreads the winds accelerate to feed the flames, a hurricane blowing inward toward ground zero. Every road is blocked, every emergency vehicle trapped or wrecked, every hospital destroyed. The fires eventually claim the lives of 195,000 people. The final toll stands at over 237,000.

In 1991 a third large ordinary chondrite, again with an orbit similar to the other two bodies of 1928 and 1989, explodes nineteen kilometers above the Sulu Sea in the Philippines with the force of twenty-seven megatons of TNT, in an area heavily used by fishing fleets from the islands of Mindanao and Palawan. Hundreds of small wooden fishing boats are set afire and dozens are torn to pieces by the blast. The final official death toll is 2,388. Because of the nature of the event, it is likely that this number is an underestimate.

SCORECARD: The number of megaton explosions in this century, ten, was very close to normal. Unfortunately, three blasts in excess of 25 megatons occurred, all of them in areas where there were many people. A quarter of a million people die in four spectacular disasters. The largest, a 103-megaton blast, occurs with an average frequency of one per two thousand years. Since we have run ten scenarios of one hundred years each, a total of one thousand years, the probability that such an event would show up in one of our runs is about 50 percent. The second-largest event, the 83-megaton airburst over France, has a mean interval of about fourteen hundred years. The third largest event in this run, the 27-megaton blast, occurs about every seven hundred years. The probability that they will all occur in the same century is of course very small—but this outcome is generated by exactly the same statistical guidelines as the other runs presented here. *In a phenomenologically complex universe, extremely improbable events are certain to happen.* Incidentally, there is nothing in the program that recognizes the existence of orbital families and clustering of impacts. Thus the three large chondritic bodies of the same composition and nearly identical entry velocities that appear in this run are the result of chance alone.

SCENARIO I

In 1903 a moderately fragile CM chondrite, a fragment from a carbonaceous near-Earth asteroid (and a possible remanent of an extinct short-period comet) enters the atmosphere at a shallow angle, passing nearly horizontally through the atmosphere over a path length of 3,000 kilometers, approaching as close as 33 kilometers above Earth's surface. Burning brighter than a full Moon, the body streaks across southeast Asia in daylight, where few notice it in a generally cloudy sky. During its lengthy pass through the atmosphere, the body loses

0.8 kilometers per second of its velocity. It emerges from the atmosphere at a speed less than Earth's escape velocity: in other words, it is captured into a temporary orbit around Earth. Such a capture event occurs only very rarely, once in several thousand tries. It requires a body that is strong enough to resist disruption by the atmosphere, and slow enough and small enough so that it can be decelerated to below escape velocity. If the body is too weak, it fragments near its point of closest approach to Earth (perigee). If it is too massive, it will not be slowed enough to be captured. If its mass is too low, it will be decelerated so much that it will fall deep into the lower atmosphere to be crushed or vaporized (or produce a meteorite fall). This is the only capture event observed in the ten century-long scenarios presented here. Two weeks later, after coasting out as far as the Moon, the body completes its long, elliptical orbit around Earth and disintegrates in the atmosphere high over the Central Pacific.

In 1923 a piece of long-period comet debris with a total energy content of thirty-three megatons enters the atmosphere over the Indian Ocean near the mouth of the Red Sea and detonates at an altitude of thirty-seven kilometers. Several tankers and a dozen freighters out to forty kilometers distance from ground zero are ignited by the radiation from the explosion. The blast wave is much too weak when it reaches sea level to blow out the flames. The death toll is 2,276, all by fire or drowning. Comets, being weak, fragment at high altitudes. Massive comets have a disturbing talent for igniting fires over great areas.

In 1989 the event of the century, a 3.7-megaton blast at twenty-one kilometers altitude, strikes off the Pacific coast of Nicaragua. One local fisherman is drowned when his boat catches fire and burns. No one connects his disappearance with the "loud thunder" heard on the mainland.

SCORECARD: Two fatal events happen in this century, one of which is "fragile," in that the only eyewitness dies. The disastrous 1923 event would very likely leave enough evidence to determine its cause. The probability of a thirty-megaton explosion in a random one-hundred-year run is about 14 percent.

SCENARIO J

In 1974, an iron meteorite showers a village near Magadan, Siberia, with shrapnel from a 370-kiloton explosion about five kilometers above the ground. Impact of the iron fragments and fires started by the explosion kills eighty-seven people. (This event is very similar to the 1947 Sikhote Alin iron-meteorite fall in eastern Siberia in our own century, except that it hits a populated area).

SCORECARD: This century registered an extraordinary number (fourteen) of megaton airbursts, of which six were over land. This century also registered the lowest fatality toll of any of the ten scenarios presented. The sole fatal event resulted from the fall of an iron, which was not even in the top twenty events by energy. Thus once again we see the poor correlation between maximum blast yield and fatality rate. Rare, strong impactors are especially hazardous.

THESE ten examples have prepared us to appreciate not only the typical impact behavior during a century, but also the vast statistical variability within any given setting. Death tolls calculated by this model usually range from zero to 300,000 per century. Several dozen one-century runs average out to about 450 deaths per year. The typical century has only one fatal event, but the size of that event can vary enormously. Many fatal events occur at times or in places where they may not be recognized as comet and asteroid explosions. Nearly half of the centuries I have run have no well-witnessed, unambiguous case of impact fatality. Since my calculations are for the present era, they assume a population density and distribution very different from that of even one hundred years ago. Allowing for far smaller past population densities, far lower seaside populations, and far less glass and gasoline in earlier times, well-witnessed lethal events should have been much less common then. The Chinese records seem to suggest a reasonable fatality rate, and lead us to wonder whether the European records of impact fatalities are as incomplete as their records of other astronomical phenomena are known to be.

Of course, ten 100-year runs give us a sample of what the most likely range of events would be in one real century. But such calculations rarely generate extremely large events. For example, a one-thousand-megaton event should occur on the average once every 15,000 years. Our 1,000 years of runs, not surprisingly, did not show a single event of this magnitude. But rare, very large events can potentially kill so many people that their *average* importance is great. Besides, a century is really a puny unit of time. The human lifespan continues to edge toward the century mark. The friends and relatives of your own personal experience, from the beloved grandparents of your youth to the grandchildren and great-grandchildren of your old age, have lives that span approximately 250 years! So let us look briefly at the results of running this same simulation program for time steps of 100 years over runs of 100 steps (10,000-year runs). As before, only the largest mass event of each time step will be reported. We have seen that the largest event is rarely the largest killer because so many of the highest-energy bodies are very fragile and explode at high altitudes. But as we move up to longer time spans and bigger objects, fragmentation in the atmosphere becomes ever less important. Projectiles that are big enough may indeed be crushed in the atmosphere, *but their pieces do not have time to get away from each other before striking the ground.* Thus for large bodies, total energy content becomes a much better yardstick of their lethality.

In ten 100 × 100 year runs, fatality rates were found to range from 720 to 6,170 per year, with an average close to 2,450. Explosions up to 8,500 megatons (8.5 gigatons), including four over 1 gigaton, were found. In the 100,000 years of runs included in this trial, there are even odds of seeing a single event larger than 16 gigatons. We saw no event this large, but did have one of 8.0 and another of 8.5 gigatons, which are roughly three times as common. Such bodies are so destructive that it is hard to find a safe place to target them on Earth. Dumping them in the ocean is no help. Consider a four-gigaton explosion in the ocean 1,000 kilometers from the nearest land: this explosion will raise an open-ocean wave 9 meters high at a range of 1,000 km. When this wave runs up onto a shore the wave height will grow by about a factor of 30 to 270 meters (high enough to crest above the eighty-sixth-floor observation deck on the Empire State Building). A 1-gigaton impact in the mid-Atlantic could devastate much of eastern North America and Western Europe. The probability of a 1-gigaton impact happening somewhere on Earth during your 250-year span of personal acquaintanceship is 2.5 percent. This is not negligible. But the hazard from tsunamis is not limited to a gigaton explosion every 10,000 years. Hundred-megaton explosions, occurring every 2,000 years, raise waves that are a third the height of those from a 1-gigaton blast. Those 100-megaton explosions, being five times as common as gigaton explosions, will typically involve one explosion that is 2.3 times less distant from the nearest seacoast than the gigaton event was. The tsunami wave that washes up on the shore nearest to the most threatening 100-megaton blast will then be 2.3/3 or 77 percent of the height of that from the 1,000-megaton explosion. When the (smaller) effects of the other five 100-megaton explosions are added in, the total hazard is closely comparable to that from the single 1,000-megaton impact. This argument cannot be extended to sizes smaller than 100 megatons because, despite the large number of 10-megaton events, few of these actually reach Earth's surface. Tsunami damage from those smaller (less than 10 megatons) impacts which reach the ocean surface can indeed occur (as in Scenarios A and F previously), but their overall lethality is much less than that of larger, rarer ocean impacts.

In general, rare, more dangerous events contribute importantly to the average death rate. *The longer the time interval we consider, the higher the average annual death rate we calculate.*

Now that we have a means of estimating these hazards, it would be a good idea to compare them with other natural environmental hazards such as storms, floods, earthquakes, and lightning, and to certain unnatural but familiar hazards such as automobile accidents, airplane crashes, and train wrecks. Clark Chapman of the Planetary Sciences Institute in Tucson, Arizona, has recently surveyed these hazards for the purpose of comparing them to the dangers posed by comets and asteroids.

Chapman makes several important points about natural hazards. First, all forms of natural disasters have "catastrophic" size distributions: a few very rare and very large events dominate the death toll. This is true of earthquakes, storms (hurricanes and typhoons; tornadoes), droughts, floods, landslides, et cetera. In a *typical* century, the death toll from earthquakes, storms, and floods will be much larger than that from impacts. But there is a very important difference between these other forms of environmental hazards and impact events: each of these hazards has a natural limit to the area (and population) that can be exposed to risk in a single event. For floods, it is the area and population of a single drainage system. This consideration makes it very unlikely that more than a few million people could ever be at risk from a single flood.

For earthquakes, the strongest possible earthquake would involve the release of all the strain energy stored on a major fault system in a single event. Many seismologists feel that this imposes a practical limit on the strength of a single earthquake event that is somewhere between Richter magnitude 9 and 10. Since most fault systems continuously relieve strain by means of frequent earthquakes much smaller than the limiting magnitude, deviant behavior building toward a giant earthquake would be easily noticed, and the dangerous accumulation of strain would make possible a prediction of the intensity (not the date) of the impending giant temblor. Evacuation of the threatened area would be possible with considerable lead time.

For storms, where the principal hazard is due to storm surges, wave action, flooding, and intense winds along coastlines, and where modern weather prediction and monitoring technologies can be brought to bear, we have already seen dramatic reductions in death tolls from even the most violent storms. Given several days of warning, protection and evacuation of threatened populations and "battening down the hatches" in the threatened area are both practical and effective countermeasures. Maximum death tolls above one hundred thousand are a thing of the past; even now, only a few areas (such as the coast of Bangladesh) remain exposed to a hazard of this magnitude. A tornado is unlikely to affect more than a few thousand people. The highest death toll ever reported is well under one thousand. Another weather phenomenon, drought, is also subject to prediction and monitoring. Enough advance warning is now available so that emergency transport of food can alleviate the situation. Where famine has been a big killer in recent decades (such as Ukraine in the 1930s and Ethiopia and Somalia in the late twentieth century), politics has invariably been a major factor in the mortality. It is hard to regard intentional political acts as "natural" disasters.

Landslides are even more local in scope. Few large cities are at hazard from this threat, and events that threaten the loss of one hundred thousand people or more are rare. Landslide-prone locales are easily identified by modern geological techniques, and relocation can be carried out where needed.

Tsunamis are a constant, low-level threat along seacoasts. The worst documented death toll for a tsunami, interestingly enough, was not from a plate-margin earthquake or continental-slope landslide, but from the sea-level explosion of Krakatau on August 27, 1883. The tsunami from that blast killed about forty thousand people. The 1815 eruption of Tambora, in Java, was even larger, but the death toll was harder to estimate. Figures of about twelve thousand are sometimes quoted, but the number is uncertain. Tsunamis caused by earthquakes are limited by the fact that earthquakes larger than Richter magnitude 10 do not occur. The largest tsunamis can only be produced by impacts.

By contrast, impact events violent enough to kill 100,000 people or more, which generally involve explosions of at least twenty megatons, occur once or twice per thousand years. The long-term average mortality rate from them in our simulations (using present-day population density and seacoast population) is about 360 people per year. On the century time-scale, tsunamis will account for less than 10 percent of the fatalities. Over time spans of thousands of years, gigaton impacts become important as sources of large tsunamis. In ten-thousand-year runs about 30 percent of the fatalities are caused by tsunamis.

Impacts are fundamentally different from terrestrial hazards in that there is no natural limit to their size. An event similar to the asteroid impact that ended the Cretaceous period would, if it occurred today, kill about 5 billion people. Since events of this magnitude occur about once every 100 million years, the average death toll from such events is only about fifty people per year. For events this large, it is not the average that matters: the event, if it were to occur, would terminate human civilization over the entire Earth. This is not an acceptable result. Unlike other natural hazards, large impacts threaten the entire planet and the very existence of civilization.

CLIMATOLOGICAL studies of the last 2 million years from Antarctic and Greenland ice-core records show that the global climate has been very unstable for that entire period of time, spanning several advances and retreats of the ice sheets and several relatively benign interglacial periods. The most astonishing feature of the entire record is that the last 6,000–10,000 years have provided by far the most stable and warm conditions in those 2 million years. *These few thousand years of exceptionally benign and stable climate coincide with the period of explosive development of human civilization.* From the birth of agriculture, the founding of the first cities and the origin of writing, through the establishment of the major religions, the birth and growth of science, technology, and industrialization, to the space age ... all this has happened in our narrow window of time. There has not been a single other time since the appearance of the earliest hominids that Earth's climate has been so conducive to civilization. We owe all human culture to a freak interval of climatic stability which we think of as "normal." We do not know why the last 2 million years were so unstable, or why the last few millennia have been so stable. We do know that civilization depends on climate stability. *When—and how—will this fool's paradise end?*

Are we at the mercy of impact-driven climate catastrophes, or is there something we can do about them? What about warning and monitoring systems to let us know what is coming? And if a body is found in a threatening orbit, could we do anything to avert an impact? These questions are addressed in chapter 15.

15

WHAT CAN WE DO ABOUT IT?

The personal safety of civilized man extends outward from the police powers in his home town to a full and vigilant patrol of outer space.

ALLAN O. KELLY AND FRANK DACHILLE
Target: Earth, 1953

A last effort was made, not indeed to turn the comet from its path—an idea conceived by that class of visionaries who recoil before nothing.

CAMILLE FLAMMARION
Fin du Monde, 1894

There is something especially horrifying about unanticipated, rare, or unfamiliar lethal hazards. Americans know that about fifty thousand people will be killed on their highways each year, and evidently discount the hazard as being so diffuse (usually only one or two people die at a time) that they need not worry about their own mortality. Certainly this knowledge does not deter them from driving vast numbers of miles each year. Some people even cultivate a disdain for safety belts, confident in their own immunity to this "normal" hazard of life. Equally certainly, the five hundred thousand American deaths each year attributed by the surgeon general and the American Cancer Society to smoking are wholly disregarded by a significant minority of the population because they occur independently, one at a time. Like traffic accidents, such events are too common to be newsworthy. But the completely trivial death toll from commercial airline crashes (zero fatalities in some years; a single fatal accident involving about one hundred people in most others) takes on a spectacular aspect precisely because the rare fatalities, when they do occur, involve one hundred or more deaths. A single airline crash that kills one hundred people is one hundred times as visible as ten thousand highway crashes that kill ten thousand people.

By any estimate, the average fatality rate from meteorite falls and impacts is equal to or greater than that from airliner crashes. But it is almost impossible to contrive a crash that kills more than 1,000 people: the most destructive airline accident ever, the March 27, 1977, collision of a KLM Boeing 747 with a Pan Am 747 on a runway (not in flight!) at Tenerife, in the Canary Islands, claimed 581 victims. On the other hand, impacts may kill thousands, millions, and even billions of people in a single event.

The psychology of "tragedies" is most interesting: when an earthquake devastates an urban area in California, causing the loss of several dozen lives, the event receives worldwide and often sensational coverage. But the earthquake's disruption of the area's transportation system forces vast numbers of people to stay off the roads for several days or weeks. The survivors take on a "disaster mentality" in which they become more cautious and more considerate of others. When they must travel, they form carpools, which greatly decreases the traffic density. The resulting saving in lives from highway accidents is often comparable to or larger than the earthquake death toll. The *net* effect of a major earthquake can sometimes be a saving of lives; but, because traffic deaths are so diffuse and the earthquake deaths can be seen as a single event, the reaction to this saving of lives is horror at the "tragedy." Public reaction to large-scale disasters bears little relation to the actual *average* numerical death toll from them. The ultimate example of "disasters" that owe their notoriety to the combination of rarity, unfamiliarity, and unexpectedness, is nuclear reactor accidents. The famous Three Mile Island disaster in Pennsylvania was comparable in its lethality to a single automobile accident of the sort that merits three column-inches in the local paper. Indeed, not a single fatality has been traced to the incident.

Clearly, the actual importance of many threats differs radically from their perceived danger. How then would people react to a natural hazard that affects the entire planet, but occurs only very infrequently? Among forest fires and brushfires, earthquakes, lightning, landslides, floods, coastal storms, tornadoes, terrestrially caused tidal waves, and impact events, the *only* one that can threaten life on a global scale, and therefore the only one that could cause the destruction of human civilization or the extinction of the human species, is a large impact. Impacts large enough to have serious global consequences (a billion casualties or more) require impactors at least several hundred meters, and probably one kilometer, in diameter. The critical explosive power for causing worldwide effects is probably close to one hundred thousand megatons (one hundred gigatons). The actual threshold may range from ten to one thousand gigatons depending upon the nature of the impactor (its strength, entry velocity, and entry angle) and where and when the explosion occurs. And of course we must make allowance for the uncertainties in our calculations of the effects of such explosions. The mean time interval between ten-gigaton events is 70,000 years; one-hundred-gigaton explosions occur at an average rate of one per 250,000 years, and one-thousand-gigaton impacts occur a little more often than once per million years. Taking your personal span of acquaintanceship, from the oldest person you knew as a child to the end of the life of the youngest person you will know in your old age, as 250 years, this means that there is a probability of 0.03% to 0.35% that someone you know will be killed in a global impact catastrophe. The probability that *you* will be killed in a global impact event is several times smaller, or 0.01% to 0.1%. Of course, if one person you know is killed in such a global catastrophe, a large fraction of your other acquaintances of the same general age will also be killed. By comparison, the probability that you will be killed in a civil airliner crash is 0.005%.

But global effects are only part of the story. We have seen that there is a wide range of lethal consequences. The least important, the death or injury of a single person struck by a falling meteorite, affects probably one to ten people per century. Villages or cities can be struck by showers of meteorites from high-altitude airbursts about once per century. Irons or other physically strong (or slow-moving) meteorites may resist atmospheric breakup to strike the surface as a single crater-forming body, or as a compact shower of iron shrapnel, with about the same frequency. Low-altitude megaton airbursts should also strike at populated areas every century or so, setting fires, shattering windows, and even demolishing buildings over an area of hundreds to thousands of square kilometers. Every three thousand to thirty thousand years a gigaton-size ocean impactor (one hundred to five hundred meters in diameter) causes a devastating tidal wave that kills many millions. And every seventy thousand to 1 million years a global billion-casualty killer will strike.

As we have found with hurricanes, predicting impact events could eliminate much of the horror and lethality associated with them. Cataloging of the orbits and properties of Earth-crossing objects is already in progress, and could be scaled up at modest expense. If we had a discovery and tracking capability, areas threatened by gigaton impactors could at the very least be evacuated. This would be ineffectual at reducing the cost of physical damage and economic dislocation, but would at least reduce the death toll to near zero. But how complete a survey do we need? That depends on the sizes and numbers of the bodies we propose to track.

The cost of finding and tracking two-thousand-plus kilometer-sized bodies that cross Earth's orbit is a few million dollars per year. Every estimate of the cost : benefit ratio that I have seen indicates that this is a wise investment. Developing a nearly complete catalog of these larger bodies is also clearly technically feasible, since such large bodies are relatively bright and relatively easily found. In fact, we have located about 10 percent of them already. In down-to-Earth terms, kilometer-sized bodies are global killers: they take the lives of a billion people per impact, and strike with explosive powers of one hundred thousand megatons at a mean rate of four impact events per million years. Thus the long-term average death rate from impacts is four billion people per million years, or four thousand people per year worldwide. The people of the United States make up about 5 percent of the global population, so the average American death rate from global-scale impacts is about two hundred per year. The death rate of American citizens from commercial aircraft crashes is one hundred people per year.

The problem with finding and tracking these very large bodies is that evacuation does not work: the effects of the disaster are global. The leading cause of death is probably famine induced by climate change. If such a body hits Earth, there are no refugia to which people can be relocated. Moving away from the computed impact area means selecting a slow death over a quick one. The death toll would be very little affected by any plausible relocation effort, since Earth's ability to support life would be universally diminished. Finding, tracking, and predicting the orbits of kilometer-sized bodies is neither technically demanding nor fiscally draining; rather, the problem arises when we ask what we would do with the knowledge. We can in fact do nothing meaningful to avoid this threat unless we use space technology to divert or destroy the threatening objects. The prospect of letting one hit our densely populated planet is unacceptable.

In our simulations, about half the fatalities are caused by smaller, much more frequent, localized events. About a quarter of the total deaths arise from tsunamis caused by impacts, and another quarter from continental cratering events and low airbursts. Meteorite falls contribute only a tiny fraction of the total. The typical tsunami event (1 gigaton; 250 meters in diameter) occurs about every ten thousand years. The population of such bodies in Earth-crossing orbits is roughly 200,000. Now it is definitely technically feasible to detect objects of this size: the Spacewatch program has found a number of near-Earth asteroids with diameters less than 10 meters. The problem is not one of sensitivity; it is one of numbers. To get thorough sky coverage requires a sizable array of telescopes. Suppose that we have a computer-driven telescope that is capable of discovering ten 250-meter asteroids per month. The cost of each such telescope is about $2 million. In order to achieve a nearly complete census of the population of near-Earth 250-meter bodies in twenty years, we need an average discovery rate of 10,000 per year, or 850 per month. Thus we require the full-time services of a network of 85 such telescopes spread around the world, or about 150 if reasonable allowance is made for observational downtime caused by cloudiness and other problems. The installation cost of the system is thus about $300 million. We should perhaps double or triple this amount to include the cost of twenty years of operations (more highly computerized observatories have lower operating costs, but cost more to install). This is still not a terrible expense: a single major unmanned spacecraft such as the US Air Force *Lacrosse* radar surveillance satellite or the *Voyager* outer-planet flyby commonly costs $1 billion.

This network of stations will discover and track our 200,000 asteroids. The survey will not be *complete* after twenty or thirty years because some of these bodies are in orbits that do not make observationally favorable passes by Earth during that time, and of course some will be missed because of poor weather or telescope downtime. Nonetheless, the survey will be more than 90 percent complete in this time, and will continue to improve with longer periods of observation. The average death rate from bodies of this size is about 1,000 per year for tsunamis and 500 per year for continental impacts. The rate of saving lives would be about 90 percent of this number (10 percent of the impactors remain undiscovered), or 1,350 people per year. With expenses of about $600 million spread over the first twenty years of intensive search, the budgetary impact is about $30 million per year, or $22,000 per life saved. This is a very reasonable cost for a life-insurance policy. But loss of life is not the only consideration. In general, the decision whether to take action against a threatening impactor would be partially based on lethality and partly based on the projected cost of the physical damage that would be done by the impact compared to the cost of taking actions to divert the body and avoid an impact.

The mean time between such impacts is thousands of years, and the time to find and catalog these bodies is decades. The probability that we will discover an object of this size on a collision course less than a year before it is due to strike Earth is about 0.01%. There is only a 2% chance that the threatening body will be discovered less than two hundred years before its impact. The most likely result is that the most threatening object discovered will not collide with us for several thousand years after its discovery. This gives us more than ample time to learn all we need to know about the body, and to develop the most appropriate technology and hardware to deal with it.

About a quarter of the total hazard is due to megaton-yield (25-meter diameter) asteroids that make airbursts at low altitudes. About once per century a megaton explosion will occur over a populated land area. The average expected death rate from airburst ignition of fires, ballistic projection of window glass, and blast-wave-induced structural failure is about one thousand people per year. Most of these fatalities (several hundred thousand) will occur in the single worst event of the millennium. There are about 20 million bodies in near-Earth orbits that have megaton impact energies. About 4 percent of the bodies in this population are physically strong irons, stony-irons, or achondrites that are capable of penetrating to the surface and excavating craters if their entry velocity is not too high. Most of these crater-forming small bodies are irons. At the opposite extreme, most of the 20 million bodies are probably structurally weak, similar to carbonaceous asteroids or cometary debris, prone to explosion at altitudes above 30 kilometers. Such explosions can be spectacular, but are not a threat to Earth's surface. The remaining 40 percent or so of the population of 25-meter bodies consists of moderate-strength chondritic asteroidal material. Slow-moving ordinary chondrite material can penetrate deep enough into the atmosphere to be a serious threat at the surface. These Tunguska-type bodies present a peculiar problem: the danger presented by them is not so easily anticipated because of the vast number of bodies in this size range that must be discovered and tracked. There are ten thousand times as many Tunguska-class (25-meter) bodies as there are kilometer-class global killers. They are so faint that they are not easy to detect. Not only are they sixteen hundred times smaller in cross-section area than the kilometer-size bodies, but there is the serious possibility that they are, on average, darker than their larger cousins. Fortunately, the ones that are hardest to find (the very dark carbonaceous and cometary bodies) are also so weak that they present a negligible threat to Earth's sur-

. . . .

face. Because of these factors, the cost of a telescope system sensitive enough to detect these bodies and extensive enough to find 20 million of them boggles the mind. Instead of the roughly 150 telescopes needed to find and track our 250-meter bodies, we would need 1.5 million telescopes (each of them superior in sensitivity, size, and cost) to find and track the 25-meter bodies. Instead of a few hundred million dollars for the entire operation, we would require an annual budget of several hundred billion dollars, comparable to the Department of Defense budget. This is not remotely feasible. The cost per life saved escalates to tens of millions of dollars per person. Clearly, political entities in the modern world do not attach anywhere near this value to the average human life. At this price, the cost of an insurance policy is prohibitive. Besides, Tunguskas are *local*, not *global*, in their effects. They do not present a hazard to civilization or to humanity, only to a single region of a thousand or so square kilometers.

I F we suppose that a search-and-tracking system is installed to find bodies down to 250 meters diameter in near-Earth space, it is likely that a number of objects in threatening orbits would be found. We have already noted that impactors in the 250-meter (gigaton explosive yield) range can be regionally devastating, but do not pose any known global-scale threat. With probable lead times of centuries, evacuation plans can easily be made and executed. The impact could then be used by fiscally conservative governments as a sort of instant urban renewal program. But with so much lead time, given the rate of advance of technology, might it not prove much less expensive and inconvenient to do something to avert the impact? If we wish to do something to the body to minimize its effects or avoid the impact altogether, what exactly should we do?

Since the time of the MIT Project Icarus report, the first idea that usually comes to people's minds when this question is raised, almost independent of their education or technical background, is, "Why not blow it up?" Clearly the prospect of a gigaton surface impact is more formidable than that of a ten-megaton impact. The latter, however terrible, is now within the realm of practical human experience; the former partakes of all those unpleasant overtones of unfamiliarity and extreme size. But supposing we split an approaching one-gigaton object into ten equal pieces, of one hundred megatons energy each, which strike Earth like a giant shotgun pattern. The radius of the

. . . .

area destroyed by an explosion is closely proportional to the cube root of the explosive yield. When you work out the arithmetic, intentional breakup looks very unwise. Ten one-hundred-megaton-surface explosions devastate an area 2.16 times as large as that destroyed by a single one-gigaton explosion with the same total yield! Splitting the one-gigaton body into a hundred ten-megaton objects causes the area of destruction to increase by a factor of 4.65. The main effect of breaking up the threatening impactor is to *increase* the damage done by a factor of several. Of course, there is more to the story than a simple comparison of surface-burst damage reveals. Suppose the gigaton surface impact was produced by an ordinary chondrite or carbonaceous chondrite body. Such materials begin to fragment in the atmosphere, but such large impactors do not have time to disperse before the entire cloud of debris slams into the ground. A smaller fragment may, however, detonate well above the ground, and not produce a crater. For example, ordinary chondrite bodies in the ten-megaton to one-hundred-megaton range tend to produce low-altitude Tunguska-like airbursts, as opposed to one-thousand-megaton bodies, which slam into the surface as a dense debris cloud. These airbursts are typically at altitudes close to the optimum burst height: they devastate an area that is about twice as large as the area that the same explosion would destroy if it were detonated on the ground! Thus the real effect of splitting a moderately strong gigaton body into ten to one hundred smaller pieces is to increase the area devastated by the impact by a factor of 4 to 10. The disruption of a threatening impactor is clearly not a sensible option unless almost all of its fragments can be diverted so as to miss Earth. But if we have the ability to divert dozens of pieces, why not elect the simpler option of gently diverting the whole thing?

The idea of diverting the course of an asteroid that is several hundred meters in diameter seems breathtakingly ambitious. Yet human mining activities routinely crush, excavate, and move comparable volumes of rock. There is an important factor that makes this scenario much less daunting: we are not trying to banish the asteroid from the inner solar system; we are merely trying to avoid a single predicted impact with Earth. Suppose our asteroid-search team finds a 250-meter body that is due to hit Earth dead center a few hundred years from now. This same body has probably been crossing Earth's orbit for 10 million to 100 million years without an impact. If we can just ease it by Earth without an impact on this one occasion, we may well buy ourselves another 30 million years to figure out what to do the next time it threatens us. So the real problem is not to devise a permanent fix: it is to avoid a specific near-term event.

We might imagine giving the asteroid a small sideways nudge so that, when it reaches Earth, it will skim by to one side of the planet rather than strike it directly. We might alternatively imagine accelerating or decelerating the asteroid along its direction of orbital motion so as to change its orbital period slightly. This would cause the asteroid to cross Earth's orbit a little ahead of or behind the impact schedule that it was following, and hence cross Earth's orbit at a point ahead of or behind Earth. The probability that we will have at least a century of advance warning is 0.99, and the probability we will have at least two hundred years of warning is 0.98: let us suppose we have two hundred years to work with. Earth's orbital velocity of 30 kilometers per second moves us along at a pretty good clip: Earth travels its own diameter (12,700 kilometers) in just 7 minutes. If the impactor was initially aimed dead center at Earth (the worst possible case), any deflection that changes the asteroid's time of crossing of Earth's orbit by more than 3.5 minutes (210 seconds) guarantees a miss. Since a typical near-Earth asteroid has an orbital period of about four years, we predict the impact about fifty asteroid-years before its occurrence. If we can change the orbital period of the asteroid by only 4.2 seconds (out of four years), the timing of the impact will be disturbed by 210 seconds and no impact will occur. This is a fractional change in the orbital period of 4.2 seconds out of 126 million seconds, a velocity change of a mere one part in 30 million. The mean orbital velocity of the asteroid is about 18 kilometers (1.8 million centimeters) per second, so the velocity change we need to produce to just barely avoid a collision is only 0.06 centimeters per second! In reality, we would want a decent safety margin, which would probably lead us to design the interceptor system to be able to divert the body by at least twice this amount. Such a small velocity change is still well below the escape velocity of the asteroid, and cannot disrupt it into several huge pieces, even if the velocity change were carried out instantaneously.

There are many methods available for making such small changes in the velocity of an asteroid. One of the favorite techniques proposed by military experts is to explode a small nuclear warhead well clear of the surface of the asteroid, perhaps two asteroid radii from the surface. But simply launching an existing ICBM at the asteroid would not work: such vehicles cannot achieve escape velocity to reach an asteroid on its orbit around the Sun. Further, missile guidance systems are designed to operate for the half-hour of an intercontinental trip, not the weeks or months required for the trip to an asteroid. The mission would have to be accomplished by a military warhead combined with a NASA planetary spacecraft bus that provides guidance and power. The spacecraft need not be massive: the nuclear explosive weighs in at only about 100 kilograms, of which only about 6 kg of bomb vapor strikes the asteroid. The shock wave from the blast is completely negligible, but the enormous thermal energy of the explosion heats a thin surface layer over the entire face of the asteroid visible to the warhead to high enough temperatures to vaporize that surface layer. The vapor departs at about 4 kilometers per second, imparting a brief impulse to the asteroid. By the nature of this technique, the force it imparts to the asteroid is very evenly distributed. Supposing that a layer 0.03 cm thick with a density of 3 grams per cubic centimeter is vaporized and departs at 4 km/s, the recoil momentum imparted to the asteroid is sufficient to change the asteroid's velocity by 10 centimeters per second. The fifty metric tons of surface material removed is more than adequate to deflect the 20 million metric tons of asteroid by the small amount required to assure a generous safety margin when it flies by Earth. The explosion of a nuclear warhead with a yield of tens of kilotons in space far from Earth guarantees that a two-gigaton explosion does not take place on Earth.

The expense of the hardware and mission operations should be around $200 million (perhaps as much as $1 billion if only one such device were built and no production-line efficiencies could be realized). Readers concerned about the environmental impact of such an explosion should realize that the asteroid would not be contaminated to any significant degree by radioactive bomb debris, since the surface layer would be boiled off by the blast. The bomb vapor would be swept out of the solar system by the solar wind at a speed of about six hundred kilometers per second. A month after the explosion the weapon debris would be 11 AU from the Sun, beyond the orbit of Saturn, and so diluted by the solar wind that it would be virtually indetectable. The one gram of matter converted to energy by the nuclear explosion would quickly get lost amid the 4 million metric tons of light given off each second by the Sun. The net result of the asteroid deflection is really a twofold benefit to Earth: a devastating impact would be avoided, and there would be one less nuclear warhead on Earth.

But suppose that we were determined not to take even the very small risk inherent in launching an inert nuclear warhead from Earth (the warhead would not be armed until it achieved escape velocity from Earth). There are several options for asteroid and comet deflection that do not involve nuclear explosions, including chemical propulsion, electrical propulsion, nuclear thermal propulsion, solar thermal propulsion, and solar sailing.

The most efficient form of chemical propulsion, burning highly volatile liquid hydrogen with liquid oxygen, releases about 10^{11} ergs per gram of propellant and produces a rocket exhaust with a speed of about 4 kilometers per second. Burning 2.5 metric tons of propellant suffices to deflect the asteroid by the minimum acceptable amount. Using a less efficient fuel mixture, such as hydrazine and nitrogen tetroxide, which can both be stored indefinitely in space, reduces the exhaust velocity to 2.5 kilometers per second and increases the required mass of propellant to 4 metric tons. The problem is simply one of landing the rocket motor gently on the surface of the asteroid and securing it to the surface in such a way that it can be fired without damage to itself or to the structural integrity of the asteroid. Neither of these seems an insuperable obstacle. Some operational considerations complicate the problem: For example, asteroids rotate with periods of about two to forty hours. Aiming the asteroid in a particular direction becomes much easier if the engine burns for a time much shorter than the rotation period, or if the impulse can be delivered at one of the rotation poles of the body. It is actually not hard to despin a small asteroid completely. A 250-meter body with a rotation period of a few hours could be completely despun by the same engine burn that is needed to deflect it. However, to do this, the engine must be securely anchored to the asteroid's equator, aimed very precisely anti-spinward, and fired tangential to the surface. Anchoring the rocket to a poorly characterized and probably very heterogeneous surface may be very difficult. "Lassoing" the asteroid and securing a cable around its equator may be the best way to grasp it firmly.

Since very long times are available for carrying out the deflection, rocket engines with very high efficiencies but low thrust levels may also be used. There is a broad class of rocket engines that derive their power not from chemical reactions but from electrical acceleration of some appropriate "working fluid" to very high speeds. These electrical propulsion devices include ion engines, arc jets, plasma jets, rail guns, and mass drivers. Such engines can achieve exhaust velocities

of ten to about one hundred kilometers per second, and therefore can achieve the same performance as a chemical rocket with much smaller expenditures of mass. Offsetting this advantage is the necessity of having a substantial source of electrical power to run the engine. That source can be either an array of photovoltaic "solar cells" that convert sunlight directly into low-voltage electricity, or a compact nuclear power source such as a radioisotope thermoelectric generator (RTG) or a small nuclear reactor. For over twenty years the Soviet Union conducted routine flights of radar ocean-surveillance satellites using Topaz nuclear reactors as their sources of power. The infamous uncontrolled reentry of the *Kosmos 954* radar surveillance satellite over the Canadian Rockies in 1979 spread a swath of radioactive fragments over thousands of square kilometers of rugged wilderness. The memory of that potentially devastating event has produced a strong negative attitude toward the use of nuclear reactors in space. The fact that the job was once done poorly means that those who know how to do it safely will be denied the opportunity. Solar cells, on the other hand, are extremely safe, but the image of a large solar cell array deployed on the surface of an asteroid next to an operating rocket engine always raises concerns that modest amounts of dust lifted by the engine might coat and shut down the solar cell array.

Two other types of engines that are independent of chemical reactions are the nuclear thermal and solar thermal propulsion systems. A nuclear thermal engine uses heat generated by a nuclear reactor to heat liquid hydrogen to temperatures much hotter than any chemical flame. The super-hot hydrogen is then vented through a rocket nozzle at speeds close to ten kilometers per second. Like electrical propulsion, this engine needs much less mass of fluid than a chemical engine. But also like an electrical engine, it needs a massive power source (the reactor). The need to launch a powerful reactor from Earth raises the same safety concerns as the nuclear electric system described above. The solar thermal engine uses a large inflatable mirror to collect sunlight and focus it upon a thrust chamber through which liquid hydrogen is pumped. The solar thermal engine is extremely safe, but the large mirror is easily distorted by gravity and hard to accommodate on the surface of an asteroid. It is also, like a solar array, vulnerable to blanketing by fine dust raised by the engine. The main advantage of the solar thermal engine is that it uses all of the incident sunlight. Even highly efficient solar cells convert only about 30 percent of the sunlight into electrical power. Thus the solar thermal engine could in principle be rather light and compact. Both nuclear thermal and solar thermal rockets achieve their greatest advantage when they use as their working fluid a material that does not have to be lifted from Earth. Perhaps the most attractive substance for use in these systems is asteroidal or cometary water. As in aikido, we can turn a resource of the threatening body against itself.

The last propulsion option is the least conventional: solar sails. A solar sail is a huge, ultralightweight mirror that runs on the recoil of sunlight reflected from its surface. No propellant or working fluid is needed. In open space not too far from the Sun, solar sails made of very thin metallic films can perform beautifully without the need to carry any propellant or working fluid. The application of solar sails to asteroid transportation is a bit less straightforward. The immense area of the sail and its extreme fragility make it vulnerable to gravitational stresses. There is also the familiar problem of how and where to attach it to the asteroid. Given time to solve them, these problems do not seem very daunting.

Several times I have commented upon hazards associated with certain spacecraft systems, especially those using nuclear warheads or nuclear reactors. When judging whether or not to use a particular system to save Earth, not one of these systems presents a hazard remotely comparable to that of an asteroid impact. If the choice were between a gigaton impact and *any* of these options, the choice would always be clear: doing *anything* (except blowing it up!) is better than suffering the impact. However, in the real world we almost always have to make choices between a variety of competing options. Given a choice between three nearly equally effective solutions, the one that offers the least potential for mischief, either inadvertent or intentional, would of course be viewed with favor. However, considering the stakes in this endeavor, it would be folly not to keep at least two different options alive lest one of them should fail for unanticipated reasons.

The choice of flight hardware for an asteroid deflection mission clearly depends on what technological options are available at the time the problem arises. In general, of course, the more different technologies we have to choose from, the more likely we are to have a good choice. It is quite impossible to guess what the preferred solution will be in A.D. 2010 or 2050, let alone 2300 or 5875. But if a threatening asteroid were discovered this year and action had to be taken in the next ten to twenty years, we would be forced to choose quickly. We almost certainly would have to make do with existing technology. There would, rightly, be very strong resistance to the adoption of any scheme that relied heavily upon untested hardware: the consequences of failure are too severe. In practical terms, then, our very limited experience with electrical propulsion systems and the fact that we have never flown a solar sail, nuclear thermal rocket, or solar thermal rocket, severely restrict our options. The fact that we have never landed a probe on a comet or asteroid leaves us woefully ignorant of the environment in which such a deflection mission would have to operate. We would, for now, be forced to rely upon nuclear warheads as the best understood and most reliable option.

All of these technologies have been under intensive study in the past. Solar sails were planned as far back as the 1960s for development for use in missions to comets and asteroids. The key mission to demonstrate solar sails was to be a rendezvous with Halley's comet in 1986. Unfortunately, NASA's reluctance to take on new technologies led ultimately to a debacle in which Japan, the European Space Agency, and the Soviet Union all launched probes to Halley while the United States stood idly by. Even worse, the ambitious American program to land a probe on a comet nucleus and fly by an asteroid at close range (the Comet Rendezvous Asteroid Flyby mission) was canceled several years ago after many millions had been spent.

Ion engines and plasma jets, extensively tested in laboratories in the 1960s, have never been integrated into the American space program, although this homegrown technology has been happily adapted by Japan for use on their communications satellites. U.S. Air Force studies of solar thermal propulsion in the 1960s eventually dried up. A handful of studies at Rocketdyne, supported mostly by internal funds, has appeared in the past twenty-five years. Nuclear thermal propulsion, the subject of an enormous Atomic Energy Commission experimental program in the late 1950s and early 1960s, was then seen as the best way to carry astronauts throughout the solar system. Nuclear upper stages were planned for the giant Saturn 5 rocket, but, when the Saturn boosters were sacrificed by NASA to make it necessary to build the space shuttle, all mention of nuclear rockets also vanished. NASA's failure to develop basic spacefaring technologies leaves us with virtually no practical short-term options. Interestingly, NASA has been extremely reluctant to fund even asteroid searches: Spacewatch has survived largely on a combination of private donations and air force money, and has in recent years been assisted by the University of Arizona/NASA Space Engineering Research Center. The latter is the only research organization in the world dedicated to the discovery, characterization, extraction, and use of nonterrestrial resources. Even the confirmation of the flux of small impactors onto Earth comes to us from the classified military Defense Support Program. Indeed, our only immediate defensive option is the one paid for by the Department of Defense: we must use nuclear weapons or nothing. Let us hope that an emergency does not arrive until we are better prepared.

O U R central conclusions, however, are unchanged by these considerations: there is no reason to panic. There is a real threat, confirmed by a wide range of evidence, but we are not helpless in the face of cosmic bombardment. We certainly have no reason to ignore the impact hazard. Once we understand that the threat exists, and once we begin to collect the information we need to deal with the threat intelligently, there is no longer any need to retreat into denial. The "giggle factor," the half-suppressed hysteria that arises from an emotional inability to deal with the truth, cannot withstand confrontation with the evidence.

First, we have the technical ability to discover and track almost every body that poses a global threat. Within a few decades a nearly complete catalog of the larger near-Earth asteroids and short-period comets could be compiled for modest cost with known technology. Certain local or regional threats, however, present problems. One obvious problem is that dangerous small bodies with high strengths, including irons, are numerous. Further, they are hidden amid an even larger population of other similar-sized bodies that present little or no threat because of their weakness. Finding, tracking, and characterizing all bodies of this size to find the relatively few strong ones is simply too vast and too expensive a task to undertake. Another problem is posed by the long-period comets. It is entirely possible that a new comet may enter the inner solar system, swing by close to the Sun, and come at Earth out of the glare of the Sun. We would not see it coming until it was far too late to do anything about it. While 99 percent of all new comets will be visible for weeks in advance, there is always the chance of one evading detection. But even the well-behaved long-period comets present a serious challenge: their orbits cannot be predicted far in advance. At the moment of discovery, a long-period comet is dashing into the inner solar system at enormous speed. If one is on a trajectory that imperils Earth, we will find out about it only a few weeks before impact. That leaves very little time for preventive measures of any kind. Fortunately, long-period comets make up a negligible percentage of the bodies that will strike Earth in the next million years. It is justifiable to ignore this minor part of the overall threat.

Secondly, massive objects found to be on threatening orbits can be deflected using techniques that combine current NASA and DoD practice, and using hardware closely related to existing military ordnance and NASA planetary spacecraft. At present, nuclear warheads exploded at a distance from the target seem the most effective and practical solution. Tom Ahrens of the Seismological Laboratory at Cal Tech has described military warheads used in this way as "weapons of mass protection." But we also know of a wide range of other space technologies that could be developed quickly to give us a variety of options for dealing with future threats. Research in basic technology could pay off handsomely.

But there is no reason to dash into an interceptor hardware-building program. Any nation that has the capability to launch a single spacecraft that can divert an asteroid and avert a multigigaton impact also has the technology to do the exact opposite. Suppose that we develop an interceptor system that has the capability of diverting such an impactor by a distance of ten Earth radii. Since near misses (within ten Earth radii) are about one hundred times as common as impacts, the opportunity to *cause* an impact arises one hundred times as frequently as the necessity to *avert* one. A one-gigaton impactor threatens Earth about once in every ten thousand years. Opportunities to cause a gigaton impact arise at the rate of about once per century. Given the political events of the last decades of the twentieth century, it would seem rash to develop a launch-ready asteroid interceptor and leave it sitting around in any country. Fortunately, the logic of the situation suggests that we do not need to have such a vehicle ready to launch.

But there is another way of looking at near-Earth bodies: a large proportion of the most threatening objects are also the most accessible bodies in the solar system for spacecraft missions from Earth. Our emphasis in this book has been on the stick rather than the carrot, but the other side of the story is equally compelling. Research at the University of Arizona Space Engineering Research Center, building upon the Spacewatch discoveries, the spectroscopic studies of NEAs by Larry Lebofsky and his coworkers, and on round-trip logistic calculations, strongly suggests that these bodies are the most promising sources of raw materials for a wide range of future space activities. They may provide the propellants for future interplanetary expeditions, the metals for construction of solar power satellites to meet Earth's energy needs in the third millennium, the life-support materials and radiation shielding to protect space colonies, and the precious and strategic metals needed by Earth's industries. The *smallest* known metallic (M-type) asteroid, 3554 Amun, is a NEA with a radius of 500 meters. It contains over $1,000 billion worth of cobalt, $1,000 billion worth of nickel, $800 billion worth of iron, and $700 billion worth of platinum metals. The total value of this single small asteroid is approximately equal to the entire national debt of the United States. By comparison, the uncontrolled impact of Amun with Earth would deliver a devastating 7-million-megaton blow to the biosphere, killing billions and doing hundreds of trillions of dollars worth of damage.

Thus we come to our final, and most startling, discovery: the stick that threatens Earth is also a carrot. Every negative incentive we have to master the impact hazard has a corresponding positive incentive to reap the bounty of mineral wealth in the would-be impactors by crushing them and bringing them back in tiny, safe packages, a few hundred metric tons at a time, for use both in space and on Earth. Remember that we will almost certainly have hundreds to thousands of years of warning time before a threatening global-scale impact. We need not be driven to rash and risky actions taken precipitously under threat of death. We will almost certainly have plenty of time to deal with the problem. This approach obviates the hazards of unauthorized deflections, since that technology would be developed only under the very improbable circumstance that a threatening object is discovered only a few decades before impact. Then, and only then, should the technology for deflection be developed, for the sufficient purpose of forestalling imminent global disaster.

Dealing with near-Earth objects should not be viewed grudgingly as a necessary expense: it is an enormously profitable investment in a limitless future; a liberation from resource shortages and limits to growth; an open door into the solar system—and beyond.

SUGGESTED READING

The reader interested in learning more about the impact hazard will find very few nontechnical sources that go into the subject in any depth, although numerous magazine articles have presented brief and fragmental overviews of parts of the problem. The best single reference for perspective on a variety of astronomical hazards is the book *Cosmic Catastrophes*, by Clark R. Chapman and David Morrison, published by Plenum Press in 1989. The treatment is much briefer and less up-to-date than the present book, but does compare impacts with other dangers such as nearby supernova explosions.

Several technical overviews of impact hazards have appeared in recent years. The earliest, devoted to an understanding of the physics of impact events and not to their effects on human populations, is *Impact Cratering: A Geological Process*, by H. J. Melosh, published by Oxford University Press in 1986. For a detailed technical assessment of the impact hazard, the only source of importance is *Hazards Due to Comets and Asteroids*, edited by Tom Gehrels, a 1994 publication of the University of Arizona Press.

There are a number of books on meteorites, of varying levels of technical detail. Perhaps the most readable is the most dated, a 1962 book by Brian Mason entitled *Meteorites*, from J. Wiley. Mason's book has long been out of print. More recent and more technical treatments include books by John T. Wasson and Robert T. Dodd, both with the same title as Mason's book. Very detailed technical information on asteroids can be found in the books *Asteroids*, edited by Tom Gehrels in 1979, and *Asteroids II*, edited by Richard P. Binzel, Tom Gehrels, and Mildred Shapley Matthews in 1989. Both books are from the University of Arizona Press. A comparable overview, *Comets*, edited by Laurel Wilkening, was issued by the same press in 1982. Dozens of popular books on comets, many lavishly illustrated, are also available.

A complementary view of near-Earth bodies, viewing them as potential resources rather than hazards, is presented in the upcoming Addison-Wesley book, *Age of Iron and Ice*, by John S. Lewis.

INDEX